Proactive Dispute Management for Construction and Related Industries

Proactive Dispute Management for Construction and Related Industries

Benedict D. Ilozor, Ph.D., MNIA, FMA, AIA Assoc., LEED AP BD+C
Eastern Michigan University
Ypsilanti, Michigan, United States

Anthony E. Sparkling, Ph.D., LEED AP BD+C
Purdue University
West Lafayette, Indiana, United States

Dennis J. Carignan
Granger Construction Company
Lansing, Michigan, United States

Srinath Perera, Ph.D., MSc IT, BSc (Hons) QS, FRSN, FAIB, MRICS, MAIQS, CQS, ICECA, MICCPM, MAIPM
Western Sydney University
Sydney, New South Wales, Australia

WILEY

This edition first published 2026

Registered Offices

John Wiley & Sons, Inc., 111 River Street, Hoboken, NJ 07030, USA
John Wiley & Sons Ltd, New Era House, 8 OldlandsWay, Bognor Regis,West Sussex, PO22 9NQ, UK

For details of our global editorial offices, customer services, and more information about Wiley products visit us at www.wiley.com.

The manufacturer's authorized representative according to the EU General Product Safety Regulation is Wiley-VCH GmbH, Boschstr. 12, 69469 Weinheim, Germany, e-mail: Product_Safety@wiley.com.

Wiley also publishes its books in a variety of electronic formats and by print-on-demand. Some content that appears in standard print versions of this book may not be available in other formats.

Library of Congress Cataloging-in-Publication Data applied for:

Hardback ISBN: 9781394282593

Cover Design: Wiley
Cover Image: Benedict Ilozor

Set in 9.5/12.5pt STIXTwoText by Straive, Chennai, India

Printed and bound by CPI Group (UK) Ltd, Croydon, CR0 4YY

C9781394282593_030726

Dedicated to global peace and harmony at the workplace and everywhere.

Contents

List of Figures

List of Tables

List of Cases from Delphi Study

Preface

Through the past and current centuries, construction and related industries have witnessed and expended enormously on disputes resulting in litigations, and deployment of alternative dispute resolution (ADR) such as arbitration, mediation, and resource-wasting private negotiations. According to 2022 research from the U.S. Chamber's Institute for Legal Reform (ILR), the costs and compensations in the U.S. tort system alone topped $529 billion. This figure represents about 2.1% of the U.S. GDP and translates to the cost of about $4,207 per American household. Add this cost to that contributed by expenditure from ADR, and the impact is staggering.

Rather than accepting these practices as the norm that we must all continue to adapt and live with, the crucial new and emerging "discipline and doctrine" of proactive dispute management (PDM) postulates that it is possible for construction project disputes to be prevented, avoided, or doused, rather than accepting and allowing them to escalate to the stage where they require investment of time and resources to resolve by conventional, time- and resource-wasting processes of negotiation, mediation, arbitration, or litigation. This book aims to offer a different approach to alternative dispute resolution (ADR) and litigation, which is evolving as PDM with three critical arms of dispute management: dispute prevention, dispute avoidance, and dispute dousing—D-PAD.

Themes Driving the Subject Matter of the Textbook

The construction industry and related sectors exist in a complex and competitive environment with many participants of conflicting interests, priorities, principles, policies, and procedures. People and companies of different capacities, views, and knowledge of the construction process work together on a project. They have their own goals and expectations, which are often discordant. Conflicts and disputes are inevitable in such a scenario. If conflicts are not proactively handled and managed, they may result in disputes, and the latter is a major factor that prevents the achievement of the triumvirate project goals of delivering on time, within budget, and to quality.

Generally, and presently, disputes in the building and construction industry tend to be addressed generically and indiscriminately from ADR to outright litigation just as any other industry. However, unlike other industries and businesses, the building and construction industry more or less uniquely does not fare as well after a few contested disputes. Some companies may not survive after dispute contests. Word gets around quickly in the industry, and before long, disputing companies are avoided like the plague by potential clients, which will ultimately lead to the decline and death of their businesses.

Due to the nature of the industry, which is anchored significantly on human relations, such connections tend to be dramatically weakened or severed by contested disputes. Hence, discerning players in the industry would prefer to never get into any contested disputes. It is not surprising to see companies in the industry retaining attorneys, not merely to deal with their contested disputes like other industries, but rather to prevent them from ever getting into any disputes that will be contested up to litigation or ADR.

Essentially, the nature of the building and construction industry calls for a different proactive approach to dispute that can save the little time they have for ADR or litigation. The approach investigated and offered in this seminal textbook is the evolving discipline and doctrine of PDM with three critical arms of dispute: prevention, avoidance, and dousing—D-PAD. By dispute prevention, we mean steps taken continually day by day to prevent disputes from arising. These terms are defined succinctly in Chapter 1.

Literally every industry professional knows what leads to disputes, or what potential areas are susceptible to disputations, or what can generate disputes. Identifying, isolating, and managing them early ahead of possible occurrence are the efficacious steps to D-PAD. This purpose is achieved through early dispute risk analysis ahead of the commencement of the construction project. A collaborative process is called for, but the importance of deploying PDM processes early cannot be overstated, much like quality assurance or control on any job. Third-party involvement, as in most ADR and litigation processes, is eliminated, as are associated unwanted exposures, expenditures, and severed human relations, as well as a possible loss of reputation and business.

Key Features and Scope of the Textbook

The key features and scope of this book cover:

1) Dispute risk analysis
2) Dispute prevention

3) Dispute avoidance
4) Dispute dousing
5) PDM deployment
6) Application of artificial intelligence in predicting disputes
7) Dispute reputation rating (DRR)

This book equips and empowers the reader with the skills to handle disputes proactively by preventing, avoiding, or dousing them. It helps the reader to avert contested disputes that are damaging to present and future business opportunities and survival. With PDM's D-PAD, third-party involvement, as in most ADR and litigation processes, is eliminated, as are associated unwanted exposures, expenditures, and severed human relations, as well as a possible loss of reputation and business.

The book is intended for both students and practitioners in construction engineering and management, architecture, ADR, and law departments. These courses, departments, schools, and professions deal with disputes, albeit in a different way from the perspective of this book, in the new and emerging discipline of PDM. Although the book advances a new discipline in dispute handling, it can be used as an extension where law and ADR are taught. A compendium is prepared and compiled here to codify and embrace PDM culture and enable it to be taught and imbibed at school and in the industry, where it should be fully deployed to change dispute outlook and handling for good. Construction and related industries should be weaned off acceptance of, and dependence on, conventional processes of negotiation, mediation, arbitration, or litigation to which they have always run to for ages to resolve preventable, avoidable, and dousable disputes.

No other textbook, discipline, or doctrine exists to compare or compete with this proposed PDM book. The closest semblance, albeit minimal, is *Dispute Avoidance* by D. Bryan Morgan[1]. While it appears to significantly share the same view on the need for dispute avoidance, it however still capitulated and conceded to ADR and such approaches by surmising thus: "It pays both to minimize the chances of them [disputes] occurring in the first place and, if that doesn't work, to entrain strategies to manage them effectively once they exist."

1 Morgan, D.B. (2008). *Dispute Avoidance: A Non-confrontational Approach to the Management of Construction Contracts*. London: RIBA Publishing.

Acknowledgments

For the foresight, sound health, determination, and power to embark on the seminal work of writing this book, the "PDM originator and lead author" immensely owes gratitude to his God Almighty who made it all possible. Next in the acknowledgments are Paul Sayer and Wiley Publishing who immediately offered the contract for this book writing following quick rave reviews from their blind peer-reviewers.

The successful writing of this textbook involved a collaboration of academic and industry professionals, thus bringing together a dynamic interchange of practical and intellectual reality that appeals to both college education and the industry at large. With much appreciation, the authors acknowledge the participation and sacrifice of scores of Granger Construction professionals in the Delphi study upon which this book was anchored. We appreciate all the photos freely available for the enhancement of the book with pictorial reliefs.

Highly acknowledged are the insights of Corinne S. Rockoff, Senior Attorney at Miller Canfield in Michigan, that helped to enrich the textbook. These early reviews helped to set the tone and the character of the textbook, making it accessible to all, irrespective of different professional backgrounds and experiences. Gratitude is due to Mathusha Francis, a Ph.D. student at the School of Engineering, Design and Built Environment, Western Sydney University, for reviews of AI-driven prediction of payment-related disputes on projects. Early review opinions and encouragement on the textbook journey also came from Dr. David Kelly, then Senior Vice President of Turner Construction and now lecturing in Civil and Environmental Engineering at the College of Engineering, University of Michigan; Dr. Ernest Uwazie, Professor of Criminal Justice, and Dan Yamshon (attorney, arbitrator, and partnering expert) both from California State University, Sacramento; Robert Chiaravalli, Esq. (a Fellow of the National Labor and Employment Relations Association, and ADR expert), Martin Gordon, Britt Howard, and the legal department of Worley. Dr. Paul Sayer must be recognized for facilitating and expediting the foresight of Wiley to offer the contract for this seminal PDM textbook writing, following quick rave reviews from their blind peer reviewers. Finally, the authors would like to thank Erica Howard, Mahi Tripathi, and Billie Johnson from Purdue University's Book Initiative editorial team for their invaluable feedback, edits, and comments that have greatly improved this textbook.

Benedict D. Ilozor (PDM originator)
Anthony E. Sparkling
Dennis J. Carignan
Srinath Perera

About the Companion Website

This book is accompanied by a companion website:

www.wiley.com/go/ilozor/PDMforconstruction1e

To access the website materials, click on the link under Related Resources on the book's page on wiley.com. You can also find this by searching www.wiley.com for the ISBN 9781394282593.

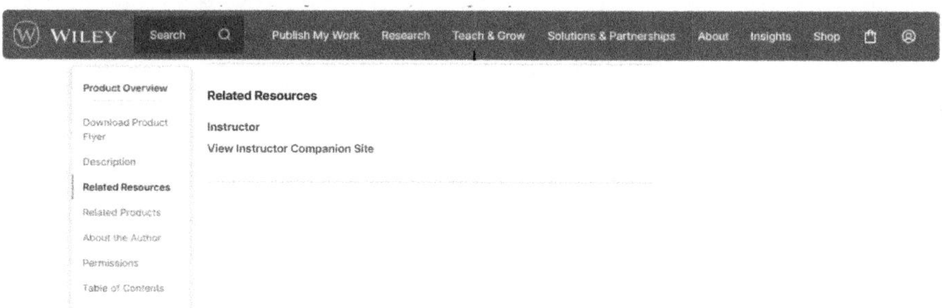

The website includes:

- Chapter review questions
- Answers

1

The Discipline and Doctrine of Proactive Dispute Management (PDM)

Readers' Learning Objectives and Outcomes

1) *Understand the discipline and doctrine of proactive dispute management (PDM) as it applies to the industry.*
2) *Define concepts of dispute prevention, avoidance, and dousing—dispute prevention, avoidance, and dousing (D-PAD).*
3) *Distinguish PDM from partnering, alliancing, integrated project delivery (IPD), alternative dispute resolution (ADR), and litigation.*
4) *Outline the common sources and triggers of disputes that can be handled with PDM in a nonconfrontational approach to project contract and delivery management.*

1.1 Introduction

The construction industry and related sectors exist in a complex and competitive environment with many participants of conflicting interests, priorities, principles, policies, and procedures. People and companies of different capacities, views, and knowledge of the construction process work together on a project. They have their own goals and expectations, which are often discordant. Attitudes among people vary as they come. Conflicts are inevitable in such a scenario, and if not proactively handled and managed effectively, they may result in disputes that can cause significant delay in the achievement of the triumvirate project goals of delivering on time, within budget, and to the expected level of quality.

There is a difference between conflict and dispute (Fenn et al. 1997), but many use them interchangeably and indiscriminately (Acharya et al. 2006; Moore 1989). Conflict applies where interests, views, and approaches are discordant and incompatible. If conflicts are not addressed and resolved, ultimately they will lead to disputes. Disputes (unresolved conflicts) can prevent the successful delivery of a project on time, within budget, and to quality. Fenn et al. (1997) went further to state that conflict can be managed, possibly to the extent of preventing a dispute resulting from the conflict, whereas a dispute is associated with distinct justiciable issues that require resolution, which lends itself to third-party intervention. In other words, disputes can be negotiated, mediated, arbitrated, or litigated, but the fact that they can be prevented, avoided, or doused (see definitions below) is not the practice of the day. They see conflict and dispute as pathological states whose causes and treatment are worth studying. They went on to ask the following questions: What then are the causes of disputes? Is it possible to identify the causal factors? If so, the construction industry's actors might seek to avoid or forestall them. However, they drove back into the usual resolution solutions—binding and nonbinding—while concluding that research into other causal factors of disputes beyond contracts may allow the development of a predictive model, allowing the parties to be aware of, and perhaps avoid, factors that cause disputes.

1.2 Distinguishing PDM from Partnering, Alliancing, IPD, ADR, and Litigation

Proactive dispute management (PDM) is a distinct, novel approach to handling disputes by aspiring ahead of conflicts and disputes to prevent, avoid, or douse them so that they are nondisruptive and either nonexistent for resolution, or don't escalate to resolution. PDM anticipates disputes and runs ahead of them by putting structures and processes in place to prevent, avoid, or douse them before they escalate and require resolution by private negotiation (relating to privity of contract between parties—owner and contractor or subcontractors and suppliers, and contractor and subcontractors and suppliers), mediation, arbitration, or litigation. By contractors, we mean general contractors (GC) or prime contractors—those who have won bids and signed contracts to take over and be responsible for overseeing projects from start to finish.

The doctrine of privity of contract in common law provides that a contract cannot confer rights or impose contractual obligations on someone who is not a part of or a party to that contract. In the privity of contract between the owner and the contractor, the word "parties" refers to the owner and the contractor. On the same hand, in the privity of contract between the contractor and the subcontractor or the supplier, the word "parties" refers to the contractor and the subcontractor or the supplier. Also, in the privity of contract between the subcontractor or the supplier and the sub-subcontractor, the word "parties" refers to the subcontractor or the supplier and the sub-subcontractor. However, we won't be considering the third privity of contract (i.e., that between the subcontractor or the supplier and the sub-subcontractor) here due to the mechanics lien law that limits lien rights to only second-tier relationships. It means that the contractor and the subcontractor can lien on the owner's project, but sub-subcontractors and others cannot. Sub-subcontractors can lien on the property of the contractor and the subcontractor, but not on the owner's property.

Generally, disputes in the building and construction industry tend to be reactively addressed through the expensive and time-consuming use of alternative dispute resolution (ADR), and sometimes litigation. Other third-party-facilitated solutions, such as partnering[1] and alliancing,[2] have been attempted, with limited success. "Lean construction"[3] and novel delivery approaches such as integrated project delivery (IPD)[4] are partly aimed at minimizing disputes despite most of their root causes being behavioral or attitudinal. Unlike other industries and businesses, the building and construction industry does not fare well after a few contested disputes, inasmuch as the industry relies heavily on the reputation of the

1 As defined by the U.S. General Services Administration, partnering is a formal management process in which all parties to a project voluntarily agree at the outset to adopt a cooperative, team-based approach to project development and problem resolution to eliminate, or at least reduce, conflicts, litigation, and claims.

2 As defined by PricewaterhouseCoopers (PwC) Australia, alliancing is a cooperative form of contracting where the participants enter into a relationship (alliance) that is designed to align the commercial interests of the participants. Each participant in the alliance will share in the success or failure of the project and in decision-making and risk management.

3 As defined by Lean Construction Institute, lean construction is a project delivery process that uses lean methods of maximizing stakeholder value while reducing waste.

4 As defined in Wikipedia, IPD is a construction project delivery method that seeks the efficiency and involvement of all participants through all phases of design, fabrication, and construction. IPD combines ideas from integrated practice and lean construction.

participants. Entities with a reputation for disputes will be avoided, which will ultimately lead to the demise of their business.

Due to the nature of the industry, anchored significantly on human relations, such connections tend to be dramatically weakened or severed by contested disputes. Thus, there is a significant interest in avoiding disputes both for the present project and for future business. It is not surprising to see the larger companies in the industry retaining attorneys, not merely to deal with their contested disputes like other industries, but rather to prevent them from ever getting into disputes that will be contested up to litigation or ADR. Essentially, ADR, litigation, and related approaches are after-the-fact damage control, or what may be termed, "medicine after death." By the time they come on board, the damage has been done, and the scars remain and build up, growing into resentment and damaging a reputation inimical to business relationships, survival, and success. While the bigger entities can afford to hire attorneys to prevent their disputes from escalating to ADR or litigation, the expense of an attorney may not be a viable option for a smaller entity that runs on tight margins.

Hence, the nature of the building and construction industry calls for a different proactive approach to dispute that can save contracting parties the little time they have for ADR or litigation, or the tedium of partnering or alliancing, and other minimally effective contractual approaches. The contractual approaches are in the mix, perhaps, because Clegg (1992) tacitly argued from a sociologist's perspective that contracts cause conflicts, and intuitively, some contracts could cause more disputes than others. In contrast, the approach offered here is the evolving discipline and doctrine of PDM using its three critical concepts of dispute prevention, avoidance, and dousing—D-PAD. PDM is augmented by predictions with artificial intelligence (AI), which will be discussed in a later chapter.

PDM is an internal conscience and constant due diligence process used to prevent, avoid, and/or douse disputes rather than waiting for them to happen and then trying to resolve. PDM is distinct from partnering or alliancing, which involves a third-party facilitator between the parties to a contract. Partnering and other solutions, such as IPD, tend to dwell heavily on contracts, which are not the only sources of disputes. Hence, the parties to the contract have to be involved in partnering, or alliancing, and IPD.

PDM here focuses principally on the contractor organization and is fluid, even though subcontractors and owner organizations can also have their own independent, internal PDM process. It takes two to tango in any dispute; when one party in a potential dispute scenario does not bring the fire (in this case, due to the contractor organization embracing PDM concepts [D-PAD] best practices ab initio), the probability of a contested dispute arising is minimal or nonexistent. While PDM considers the totality of the project, including its owner, it does not force other project participants and stakeholders to engage in team integration

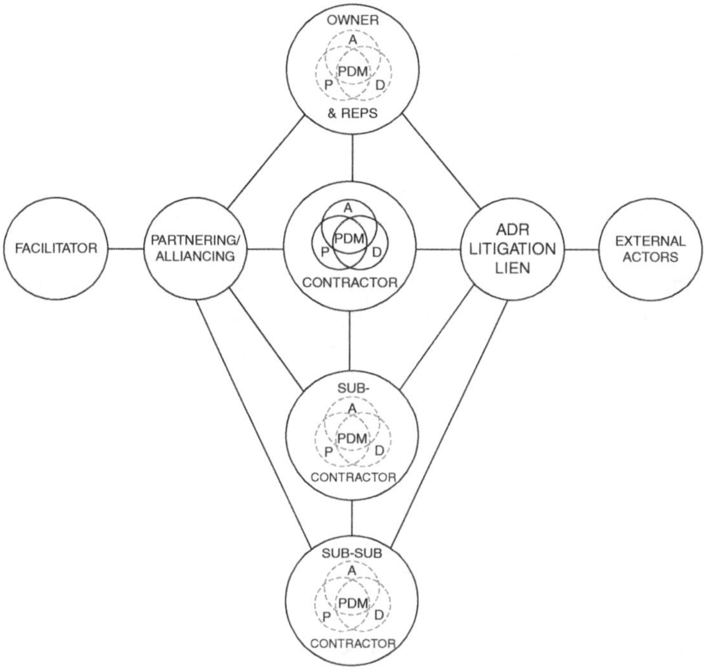

Figure 1.1 Proactive dispute management (PDM) ecosystem. *Source:* Ilozor (2019).

practices often codified in contracts and associated approaches, such as IPD, partnering, or alliancing. Rather, PDM is an internal philosophy and a paradigm shift that is holistically practiced by individuals and organizations to proactively moderate and mitigate disputes from arising. However, emphasis in this book is central on the contractor organization, as its personnel have the greatest opportunity to make an impact on the dispute climate of a project (Diekman et al. 1994). Hereunder is an illustration of PDM's ecosystem from the lead author's collection of illustrations (Figure 1.1).

1.3 Defining the Three Critical Concepts of PDM–D-PAD

The three critical PDM concepts are as follows:

1) Dispute prevention is an organizational cultural initiative that refers to the steps that are consciously and continuously taken by an entity to prevent disputes from arising. Dispute prevention strategies include a strategic or longer-term culture of working with comprehensive planning and detailed

design, ample and effective communication, unambiguous and explicit contracts, proactive risk analysis and assessments of projects, imbibing of total quality management (TQM), quality assurance (QA), and quality control (QC) principles, six sigma principles, dispute reputation rating (DRR), openness to collaborative approach, and effective document handling, distribution, and interpretation. A culture of thorough value engineering, and especially project constructability analysis to hash out and clear all construction bottlenecks, is a potent measure in the dispute prevention toolbox.

2) Dispute avoidance, taking the elephant in the room analogy, refers to actions to keep or navigate away from extant potential dispute flashpoints and not do anything to agitate them in the project delivery process. Due to the ever-presence of Murphy's law, even with our best intentions and efforts, things that can go wrong will go wrong. Dispute avoidance strategies include both a priori and a posteriori knowledge and experience in the management of the construction process, such as cement curing and setting, that may delay some sequence of the schedule and the works. Projecting and envisioning such potentially disruptive aspects of the project enables the ability to navigate around them and keep the work moving forward with minimal or no disruptions, which can avoid conflicts and disputes. Again, planning ahead and thorough prior analysis of the project for such risks are key to avoiding consequential conflicts and disputes from them. Hence, Disputes Potential Index (DPI)—as piloted by a study sponsored by Construction Industry Institute's (CII) Disputes Avoidance and Resolution Task Force to predict contract disputes (Diekman et al. 1994)—and other dispute-predicting tools and technologies can be essential measures in the dispute avoidance toolbox.
3) Dispute dousing is engaged when dispute prevention and dispute avoidance fail, and it involves the deployment of quick actions to neutralize and extinguish emergent disputes before they deteriorate into costly and time-consuming ADR or litigation. Dispute dousing can be achieved by continuous close monitoring of the process to ensure early identification of budding dispute issues, open communication channels with the parties and participants, encouragement of collaborative solutions to emerging issues, and clear task and responsibility assignments and leadership. Charisma and probably some conflict handling and resolution training—as well as intermittent and regular audits and reviews—are helpful within the dispute dousing toolbox.

Most experienced industry professionals are aware of what leads to disputes, or what potential triggers and areas are susceptible to disputes, or what can generate them. It means that most of these disputes are predictable, and measures can be put in place to prevent or avoid them, and if both steps fail, then swift action is taken to douse them quickly before they escalate to a degree that requires resource-consuming resolution. Hence, identifying, isolating, and managing them early, ahead of possible occurrence, are the efficacious steps to D-PAD. This purpose is achieved through early internal dispute risk analysis before commencement of the construction project. Much like TQM, QA, and QC on any job, engaging in an internal PDM process prior to construction can avoid third-party involvement such as partnering or alliancing, ADR, and litigation, thus eliminating expense, lost time, owner involvement, and risk of loss of reputation and future business.

Besides the dispute types offered under Table 12.1, one of the most comprehensive categorizations of common causes of disputes was given by Cakmak and Cakmak (2014) as extracted in Table 1.1. PDM through D-PAD cannot be effectively deployed without appreciating where and how these common disputes originate.

Table 1.1 Common causes of disputes by categories.

Category of disputes	Causes of disputes
Owner related (A)	1. Variations initiated by the owner (A1) 2. Change of scope (A2) 3. Late giving of possession (A3) 4. Acceleration (A4) 5. Unrealistic expectations (A5) 6. Payment delays (A6)
Contractor related (B)	1. Delays in work progress (B1) 2. Time extensions (B2) 3. Financial failure of the contractor (B3) 4. Technical inadequacy of the contractor (B4) 5. Tendering (B5) 6. Quality of works (B6)
Design related (C)	1. Inadequate/incomplete specifications (C2) 2. Quality of design (C3) 3. Availability of information (C4)
Contract related (D)	1. Ambiguities in contract documents (D1) 2. Different interpretations of the Contract provisions (D2) 3. Risk allocation (D3) 4. Other contractual problems (D4)
Human behavior related (E)	1. Adversarial/controversial culture (E1) 2. Lack of communication (E2) 3. Lack of team spirit (E3)
Project related (F)	1. Site conditions (F1) 2. Unforeseen changes (F2)
External factors (G)	1. Weather (G1) 2. Legal and economic factors (G2) 3. Fragmented structure of the sector (G3)

Source: Cakmak and Cakmak (2014) / with permission of Elsevier.

Table 1.2 Conflict and dispute in construction from the literature.

Research	Sources of disputes from the literature
Bristow and Vasilopoulous (1995)	Five areas: unrealistic expectations; contract documents; communications; lack of team spirit; and changes
Conlin et al. (1996)	Six areas: payment; performance; delay; negligence; quality; and administration
Diekmann et al. (1994)	Three areas: people; process; and product
Heath et al. (1994)	Seven areas: contract terms; payment; variations; time; nomination; renomination; and information
Hewit (1991)	Six areas: change of scope; change conditions; delay; disruption; acceleration; and termination
Kumaraswamy (1996)	Two areas: root causes and proximate causes
Rhys Jones (1994)	10 areas: management; culture; communications; design; economics; tendering pressures; law; unrealistic expectations; contracts; and workmanship
Semple et al. (1994)	Four areas: acceleration; access; weather; and changes
Sykes (1996)	Two areas: misunderstandings and unpredictability

Source: Fenn et al. (1997) / with permission of Taylor & Francis.

Fenn et al. (1997) provided a literature collection of conflict and dispute causes in construction, as shown in Table 1.2.

1.4 Conclusion

Contractual parties' desire to avoid litigation from disputes has pushed them toward different deliveries and contracts besides the lump-sum approach. Innovations in construction delivery processes and methods have seen more collaborations between designers and builders, as well as all participants and stakeholders. Despite these new welcome developments and the availability of many dispute minimization and long-practiced resolution methods in construction and related industries, disputes between parties still remain of great concern to the participants and stakeholders. Conflicts and disputes abound where two or more parties have to work together on a project or projects, and their potential rises even more in a field such as construction and related industries with many players and parties on a common situs. The choice is ours as to how to handle or manage these conflicts and disputes: either reactively and through third-party and contractual involvement, as in the case of ADR, litigation, partnering/alliancing, and IPD; or proactively, as with PDM concepts such as D-PAD. Reactive, contractual, and third-party-moderated approaches have been adopted as the norm over many years, and we have accepted them. However, we don't have to live with them anymore with the PDM concept now available, as continuing with the old ways will continue to be damaging in terms of time and cost, and will damage relationships, reputation, and competitiveness.

The opportunities in construction are limitless with the advancement of time and technology. The industry spans from residential projects to commercial, industrial, and civil/heavy/highway engineering projects, which have all grown in size and complexity. Many people work in the industry, and with this scenario comes the potential for conflicts and disputes. Managing conflicts and disputes so that the business of construction is not disrupted or shortchanged is of great benefit to all parties and participants. PDM provides new solutions unencumbered by contracts, onerous processes, and third-party interventions.

In summary, the construction industry builds the nation, making up as much as 10% of some nations' gross domestic product (GDP). Many parties and participants from the construction industry are involved in project delivery, including owners; professionals like architects, engineers, and construction managers; contractors and the trades (subcontractors, sub-subcontractors, suppliers, and vendors); and regulatory authorities. This scenario breeds the potential for conflicts and disputes. When two or more are gathered together, or working together, the probability of conflicts and disputes rises. The participants and players portend their own conflicts and disputes. Knowing about the types of conflicts and disputes associated with each participant and situation is important, as are the means to handle or resolve them. However, far more important and profitable is recognizing the flashpoints and triggers of conflicts and disputes beforehand, and figuring out a way to prevent, avoid, or douse them before they flare up and cause disruption to work, or cost time and money, or cause disaffection among the parties. We will look at the dispute flashpoints and triggers associated with different participants, players, and situations through the lens of PDM. This approach allows us to identify ways to prevent, avoid, or douse these triggers from the contractor's perspective, thereby saving the contractor's time and money. Furthermore, it helps prevent strained business relationships and loss of competitiveness that can arise as a consequence of even the best resolved disputes. The dispute flashpoints and triggers calling for PDM will be presented here in three chapters, namely:

Chapter 2: Owner-related dispute flashpoints,
Chapter 3: Contractor-related dispute flashpoints, and
Chapter 4: Miscellaneous dispute flashpoints that include:

a) Subcontractor and supplier-related dispute flashpoints,
b) Construction documents-related dispute flashpoint,
c) Contract-related dispute flashpoints,

d) Attitude and disposition-related dispute flashpoints,
e) Nature and uniqueness of project-related dispute flashpoints, and
f) External factors-related dispute flashpoints.

1.5 Case

Dispute Dousing:

1) *Case shared by Granger's President, Dennis Carignan*: We were selected to construct a student recreational facility to be paid for via a student fee that was being charged immediately (before the facility was available for use). This reality created a strong sense of urgency to get the facility designed and constructed. We were hired after the design team, primarily because we demonstrated a sense of urgency and had a good track record of delivering on our promises for fast-track projects. Once hired, we began pulling early work ahead to both show progress and create a more realistic construction duration and still achieve the desired completion date. The entire team was clear about the plan and committed to the dates. We worked intensely with the design team. Site preparations, under-slab utilities, building foundations, and slab on grade were all pulled ahead of winter conditions, allowing for steel erection to take place through the winter months. Unfortunately, some combination of limited resources, competing priorities, some envelope changes, and a bit too much optimism contributed to a slew of structural steel RFIs, which delayed delivery by nearly two months. Given the political nature of the project, a stern letter of concern was sent to both the designer and the builder (same letter). It is very tempting to formally defend your position when this happens. After a brief discussion, we realized the predicament that the administration was in and decided that a rebuttal of facts, no matter how true they were, was not going to improve the situation. Administration wanted to feel heard, and calling out project partners would only drive a wedge in the team at a time when we needed even more cooperation. We instead met with the designers and key trades, agreed on a joint response with assurances that we understood, had analyzed the contributing factors, that we had a very good and clear path forward to prevent any further delays, and reiterated our commitment to the completion date. Though it was not easy to swallow our pride in the moment, we reinforced team trust with our partners, satisfied the owner's concerns, and went on to fully recover the lost time through adjusted details, sequencing of the work, and minor selective overtime. We consider this team-first philosophy essential in proactive dispute management.
2) *Case shared by Granger's President, Dennis Carignan*: The second example is the occupied/phased extension of a dental school. This was a $125M project for

a university. Our scope was to completely replace the electrical and mechanical systems for the entire building while sequentially renovating each space and maintaining egress, wayfinding, temperature, and dust control, and maintaining noise levels to accommodate ongoing patient care, research activities, and teaching class schedule. Each area's schedule was critical to be completed before the occupants could relocate and the next phase could begin. Many of the phase start dates were linked to the academic calendar, so missing them by even a week was not an option. The risk of additive/cumulative delays was extremely high. During the first phase (main infrastructure replacements that allow future phases to commence), previously concealed existing conditions were found to be substantially different from the design documents. Time lost to resolving discrepancies, adjusting the scope, and resequencing activities all had to be made up via overtime and shift work during an extremely tight labor market. Tensions ran extremely high for far too long. The management team agreed to meet the needs and reengineered meeting rhythms and agendas. Monthly team culture meetings were scheduled to vent frustrations to/with each other and agree on adjustments and renew commitments to each other. The as-built conditions continued to vary from those that were anticipated, which required a few interim milestone date adjustments, but in the end, the project finished exactly on its original substantial completion certificate. Through $24M of change management over four years of extraordinarily tense conditions, not a single subcontractor claim had to be escalated to any formal ADR measures. Even more impressive was that despite the difficult working conditions that appeared to have no end in sight for years on end, not a single Granger employee on that project exited the company during the project. The project went on to win a Build Michigan and a Build America Award for its perseverance in overcoming so many challenges to reach a successful outcome.

3) *Case shared by Anthony Sparkling, Ph.D. (former Granger Project Manager)*: The project was a new $100M enterprise data center with myriad stakeholders and project participants. It was delivered using Design-Build (DB) with early specialty contractor (i.e., electrical and mechanical contractors) involvement. Due to the complexity of the build, the owner hired an experienced consultant to manage quality, drive decisions, and to challenge the design—their primary function being to ensure that the design and construction team were working assiduously to achieve Uptime Tier 3 certification (i.e., a data center with multiple paths for power and cooling and systems in place to update and maintain it without taking it offline—concurrent maintainability). These types of data centers are designed with an expected uptime of 99.982% (1.6 hours of downtime annually).

As the project underwent final commissioning (Cx) and approached owner turnover, unexpected flashpoints would arise and were quickly doused. One such instance involved a bypass pipe that needed to be installed between two chilled water loops feeding the data hall's cooling systems to prevent stagnation from occurring in the normally closed tie pipe. The mechanical engineer provided an architect's supplemental instruction (ASI) with enough detail for the mechanical contractor to complete the additional work. After completing the work, the consultant challenged the effectiveness of the installation and its adherence to the design intent.

To avoid escalation and additional rework costs, we decided to bring all the relevant stakeholders to the field to discuss the situation. The owner's team, including the consultant, end-user groups, project management staff, and the facility operators, met with the contractor team (i.e., GC, mechanical subcontractor, and design engineers). Once we started reviewing the installation and discussing the work, the situation immediately became contentious. I asked everyone to stop for one minute so we could approach things objectively and allow everyone to add shared knowledge to the discussion. I worked my way around to each of the stakeholders that would be affected by the current installation or involved in rework. With clear consensus and agreement from the design engineer and construction team, maintaining that the installation met the design intent, the facility operators lauded how they thought the installation was better than the changes proposed by the consultant. They went on to say that "the design exceeded the requirements of their internal design specifications for small pipe sections." After these last comments, the consultant acquiesced to the will of the majority and the core owner team, who suggested that we move on.

And while countless flashpoints like this occurred throughout the final Cx and closeout phases, none of them led to arbitration or litigation. With diligence and our attentiveness to the owners' concerns, we were awarded their redundant $140M enterprise data center project that was continuously being designed and budgeted for via a change order.

References and Suggested Further Reading

Acharya, N.K., Lee, Y.D., and Im, H.M. (2006). Conflicting factors in construction projects: Korean perspective. *Journal of Engineering, Construction and Architectural Management* 13 (6): 543–566.

Bristow, D. and Vasilopoulos, R. (1995). The new CCDC 2: facilitating dispute resolution of construction projects. *Construction Law Journal* 11 (2): 95–117.

Cakmak, E. and Cakmak, P.I. (2014). An analysis of causes of disputes in the construction industry using analytical network process. In *Second World Conference on Business, Economics and Management – WCBEM 2013. Procedia – Social and Behavioral Sciences* 109: 183–187.

Clegg, S. (1992). Contracts cause conflict. In: *Construction Conflict: Management and Resolution* (ed. P. Fenn and R. Gameson), 128–144. London: Chapman & Hall.

Conlin, J., Langford, D., and Kennedy, P. (1996). The relationship between construction procurement strategies and construction contract disputes. *CIB W92, North Meets South*, Durban, South Africa, 14–17 January 1996.

Diekmann, J., Girard, M., and Abdul-Hadi, N. (1994). Disputes potential index: a study into the predictability of contract disputes. Construction Industry Institute, Source Document 101, The University of Texas, Austin, TX.

Fenn, P., Lowe, D., and Speck, C. (1997). Conflict and dispute in construction. *Construction Management and Economics* 15 (6): 513–518. https://doi.org/10.1080/014461997372719.

Heath, B., Hills, B., and Berry, M. (1994). The origin of conflict within the construction process. *Proceedings of the First Plenary Meeting of TG15*, Publication 171. Rotterdam, The Netherlands: CIB.

Hewit, J. (1991). *Winning Construction Disputes – Strategic Planning for Major Litigation*. London: Ernst and Young.

Ilozor, B.D. (2019) Dispute prevention, avoidance, and dousing – DPAD: A discipline and philosophy of proactive dispute management. Unpublished manual presented to Vice President & General Manager, Turner Construction Company at Eastern Michigan University, USA.

Kumaraswamy, M. (1996). Conflict, claims and disputes in construction. *Engineering and Architectural Management* 4 (2): 95–111.

Moore, C. (1989). *The Mediation Process*. San Francisco, CA: Jossey Bass.

Rhys Jones, S. (1994). How constructive is construction law? *Construction Law Journal* 10 (1): 28–38.

Semple, C., Hartman, F., and Jergas, G. (1994). Construction claims and disputes: causes and cost/time overruns. *Journal of Construction Engineering and Management ASCE* 120 (4): 785–795.

Sykes, J. (1996). Claims and disputes in construction. *Construction Law Journal* 12 (1): 3–13.

Chapter Review Questions

1) Which of the following statements is *true* about PDM?
   a) It is an in-house dispute handling process.
   b) It is a dispute resolution method.
   c) An independent third-party facilitates the process and ensures equitable dispute handling.
   d) It has been in use in the industry for as long as ADR, IPD, partnering, and alliancing.

   e) All of the above.
   f) None of the above.
2) The potential for conflicts and disputes in the industry will reduce with the advancement of time, innovations, technology, and collaborations. True/False?
3) Conflicts and disputes are the same and can be used interchangeably. True/False?
4) Which dispute handling and minimizing method(s) is/are nonreactive and doesn't/don't require third parties? (Choose the right answer[s])
   a) Litigation
   b) IPD
   c) ADR
   d) PDM
   e) Partnering
   f) Alliancing
5) PDM application is binding on the parties to a contract to prevent, avoid, and douse disputes before they escalate to the need for resolution. True/False?
6) Resolving disputes tends to strengthen human relations and promote business growth and success among participants and parties of the construction and related industries. True/False?
7) A formal management process in which all parties to a project voluntarily agree at the outset to adopt a cooperative, team-based approach to project development and problem resolution to eliminate, or at least reduce, conflicts, litigation, and claims. This statement refers to ________________.
   a) Litigation
   b) IPD
   c) ADR
   d) PDM
   e) Partnering
   f) Alliancing
8) ________________ is a construction project delivery method that seeks the efficiency and involvement of all participants through all phases of design, fabrication, and construction.
   a) Litigation
   b) IPD
   c) ADR
   d) PDM
   e) Partnering
   f) Alliancing

9) ________________ is a cooperative form of contracting where the participants enter into a relationship that is designed to align the commercial interests of the participants.
   a) Litigation
   b) IPD
   c) ADR
   d) PDM
   e) Partnering
   f) Alliancing
10) A business with a reputation for well-resolved disputes (low DRR) will be successful among competitors. True/False?
11) PDM is applicable only to general contractors of the construction industry. True/False?
12) PDM concepts include ______________.
   a) Dispute prevention
   b) Dispute avoidance
   c) Dispute dousing
   d) All of the above
   e) None of the above
13) ______________ refers to the steps that are consciously and continuously taken by an entity to prevent disputes from arising.
   a) Dispute prevention
   b) Dispute avoidance
   c) Dispute dousing
   d) All of the above
   e) None of the above
14) ______________ refers to actions to keep or navigate away from extant potential dispute flashpoints and not do anything to agitate them in the project delivery process.
   a) Dispute prevention
   b) Dispute avoidance
   c) Dispute dousing
   d) All of the above
   e) None of the above
15) ______________ refers to the deployment of quick actions to neutralize and extinguish emergent disputes before they deteriorate into costly and time-consuming ADR or litigation.
   a) Dispute prevention
   b) Dispute avoidance
   c) Dispute dousing
   d) All of the above
   e) None of the above

16) PDM can be applicable to other industries outside of construction. True/False?
17) Which dispute handling process(es) may not require third-party involvement? (Choose the best answer[s])
    a) Litigation
    b) IPD
    c) ADR
    d) PDM
    e) Partnering
    f) Alliancing
18) Which dispute handling process(es) may be unconnected to the contracts between the parties? (Choose the best answer[s])
    a) Litigation
    b) IPD
    c) ADR
    d) PDM
    e) Partnering
    f) Alliancing
19) Contracts cause disputes, as argued by a sociologist, Clegg, in 1992. True/False?
20) Some contracts could cause more disputes than others. True/False?

Exercises

Exercise 1.1 Identify a contractor or an owner in your community and inquire about any past projects; discuss dispute issues that happened and how they were handled effectively or otherwise. Ask if they think that any/some/all of the disputes could have been better handled differently and how.

Exercise 1.2 Research prominent past projects in which there was litigation in the news, and articulate the disputes that led to litigation resolution.

Discussion

1) When compared to other fields, construction and related industries are prone to conflicts and disputes. Do you agree or disagree? Discuss why on either side of the argument.
2) What attributes of a business that earn a good DRR are deemed favorable to owners and ensure its competitiveness?

2

Owner-Related Dispute Flashpoints

Readers' Learning Objectives and Outcomes

1) *Analyze the various owner-related flashpoints and triggers of conflicts and disputes in the industry.*
2) *Conceptualize plausible prevention, avoidance, or dousing approaches that may be applicable to these conflicts' and disputes' flashpoints and triggers.*
3) *Suggest prevention, avoidance, or dousing strategies effective or suitable to each conflict's and dispute's flashpoints and triggers.*

2.1 Introduction

Owner dispute flashpoints and triggers can come from several scenarios, such as addenda, site conditions, change orders (variation orders, as they are called in most English-speaking countries), scope creep, turnover or handover hitches, constructive acceleration, lofty and unrealistic expectations, delivery delays of lead items or provisional and prime cost sums, and payment delays.

2.2 Addenda

Addenda or addendums are necessary or precautionary corrective actions to clarify, revise, or supplement the documents of a contract. During the bidding phase of projects, it may be discovered that there are omissions or aspects of the work or documents to be clarified, corrected, revised, or supplemented. This leads to something extra being added, and that is called an addendum for one, and addenda or addendums if more than one. While addenda are helpful to the completeness of the documents of a contract, they are a means to ensure fairness and transparency to all bidders, and help to avoid later misunderstandings and disputes during construction, they can sometimes signal and set an early ominous tone for the smooth implementation of the contract. Where you have many of them, they may signal and raise anticipation or apprehension for more omissions and commissions to deal with through many requests for information (RFI) and change orders. Omissions (failing to do the right things) or commissions (doing things wrongly) will result in undesirable dispute outcomes or the potential for such outcomes. The problem with too many addenda is their propensity to significantly modify the original contract even before construction commencement.

Here are a few reasons why addenda may have the potential to lead to disputes:

1) *Communication timing*: If issued late in the bidding process, or if there are communication issues with their distribution to the bidders, some may not receive the updated, corrected, or supplemental information on time before bid closing. This untimeliness may lead to confusion and potential disputes over which version of the documents is valid to follow.
2) *Scope changes or scope creep*: Sometimes, addenda may lead to significant changes to the scope and requirements of the project, for which some bidders might be ill-prepared if they win the bid and these addenda are not clearly communicated, or they may impact the cost and time of the project. The bid winner or the contractor may struggle with the changed scope, cost, and timeline, and the potential to argue and cut corners or seek extra compensations increases, as does the probability of contested disputes.

3) *Different interpretations of the addenda*: Despite the fair and good intentions of the addenda, they are still subject to different interpretations by owners and the bidders, however so slight. Disputes can arise if the bidders or contractors hold different interpretations of the intent and language of the addenda.
4) *Addenda conflicting with other provisions of the contract*: In some cases, addenda may conflict with other provisions of the contract by introducing ambiguity or inconsistency, leading to potential disputes if and when the contractor seeks adjustments to the project's scope, compensation, and schedule.
5) *Discrepancies introduced by the addenda*: Addenda can sometimes muddy the documents and introduce discrepancies in the form of errors and omissions. Contractors may seek further clarifications, and this can take time and delay the construction process, and may potentially lead to disputes.

The question is therefore whether potential disputes from addenda can be prevented, avoided, or doused. Preventing them appears far more difficult, if not impossible, than avoiding or dousing them. The reason is that the mere fact of addenda being in the process portends the presence of these triggers, latent or otherwise. Avoiding them seems more plausible as addenda originate principally from owners and their representatives and are as varied as they come. One suggestion to avoid disputes from addenda is to exhaustively X-ray them from the start by thorough clarifications and documentation, effective change management, and clear and comprehensive communications with the owners and their representatives, including the design team. The potential or embodied dispute-risk analysis (DRA) should be conducted to determine how they can be avoided, and if disputes insist on flaring up, to douse them swiftly and effectively before they escalate. One might ponder how the inclusion of addenda may have significantly modified the terms and conditions of the original contract and the project scope; hence, they should be envisioned, discussed, and agreed upon ahead of signing the contract.

2.3 Site Conditions

Site conditions refer to the physical and environmental realities of the location where the project will be situated. These conditions affect the planning, design, and construction processes of the project. Site conditions include:

1) *Topography*: The contour of the site (slopes, valleys, hills, etc.) can influence the layout of the project, drainage patterns, landscaping, extent of earthwork, soil stabilization, and shoring by retaining walls and other means.
2) *Nature and type of soil*: Understanding the properties of the soil—soil type (loam, humus, clay, sand, alluvial, and gravel), water retentivity, and bearing

capacities—is fundamental to determining the type and complexity of foundations, excavation methods, soil stabilization, and shoring that would be required for the project.

3) *Geotechnical conditions*: A site's geotechnical factors, such as underground water table, swamp, potential for landslides, and seismic activities, can affect its stability and safety for the project.
4) *Prevailing weather and climate*: Factors such as climate zone, temperature, humidity, precipitation, and wind speed can affect a project's schedule, material performance, labor and equipment productivity, and safety. Extreme weather conditions may call for special construction and protective measures: for instance, building in winter in a temperate climate may require a different type of cement and additives.
5) *Applicable regulations*: There may be local, state, or federal regulations pertaining to the site with regard to various environmental protections for flora and fauna conservation, or cultural and historical heritage protection. Complying with these requirements is essential and may require environmental assessments, permits, and inspections, all of which can affect the construction process and its schedule and progress.
6) *Site utilities and services*: Existing services and infrastructural utilities such as water, sewer, gas, electricity, and telecommunication facilities must be ascertained. Timely coordination with providers to ensure proper connections helps to avoid delays and disruptions. For instance, timely calling 811 (known as Miss Dig in Michigan) helps to avoid tampering with underground utilities and services that will cost and delay the project.
7) *Accessibility and proximity*: Project delivery efficiency, material delivery schedules, and logistics planning are all affected by proximity to labor and material supplies and the availability of transportation systems such as access roads, railways, waterways, and airports.
8) *Locational constraints*: This has to do with zoning restraints, building codes, easements, and property boundaries dictating the permitted uses of the site and the design and layout of any project on it. Compliance with these restrictions is crucial to avoiding delays from jurisdictional authorities' intrusions and disruptions of the project, including work stoppage, fines, and litigation.
9) *Cultural and historical aspects*: Projects on sites of cultural, archeological, heritage, or historical significance may require special assessments and inspections by preservation authorities in order to ensure respect and protection of these assets.
10) *Contiguity and vicinity*: Immediate neighbors, community, and public amenities must be recognized and respected to minimize disruptions, limit noise, dust, and debris, as well as traffic, to maintain harmony with these stakeholders.

Evaluating, understanding, and effectively handling the project's site conditions are critical to its successful delivery within budget, time, quality, and possibly safety. Comprehensive feasibility studies and site assessments, together with coordination with experts and authorities, are crucial in mitigating risks from site conditions and optimizing project outcomes and success.

Site conditions can generate enormous potential for disputes, and here are a few reasons for this scenario:

1) *Hidden and unidentified underground conditions*: Most contracts stipulate that the owner must guarantee that the site will support the project, including both seen and unforeseen realities of the site, and what is above ground and what is underground. Despite all prior geotechnical boring tests, site assessments, and investigations, unforeseen conditions may still arise, such as contaminated soil and other soil-related environmental hazards, rocks and boulders, unmarked underground services, and utilities. Encountering such can lead to unexpected disputes because of their impact on costs, schedule, and scope, as well as the party taking the responsibilities and bearing the costs and consequences.
2) *Contract clause covering differing site conditions*: While some contracts may come with a clause to cover differing site conditions, because they can come in different sizes and scopes, the potential for disputes still exists with regard to the interpretation and application of this clause. Differing site conditions can quickly modify the contract scope, cost, schedule, and sequence; with accompanying consequential damages, claims may then become potential issues of dispute. Contaminated soil may require special handling and remediation, including the engagement of environmentalists and specialists, and deployment of specialized techniques and equipment for which the contractor is unprepared or ill-equipped.
3) *Discharge of contractual obligations*: The same contracts charge that the contractors exert due care and diligence in inspecting the site, ascertaining and clarifying any discrepancies with the owner ahead of construction commencement. Due to this clause, it would amount to a breach of contract for the contractor to stop any aspect of the project due to the differing site conditions that have added to the scope. This ambivalence can be a source and trigger of dispute.
4) *Cursory and casual site investigations and assessments*: The contractors may be ill-equipped to conduct thorough site investigations and assessments, and it is understandable, as they are uncertain of winning the bid among other bidders. While the contractors can argue that they relied on the site information provided by the owner, some owners may claim that the contractors ought to have identified and accommodated the differing site conditions ahead of submitting their bids. This dichotomy of perspectives can be a source and trigger of dispute.

In order to eliminate or mitigate disputes arising from site conditions, parties to the contract need to conduct thorough site investigations and assessments, assess the risks of unforeseen conditions, and marshal out clear contractual obligations, responsibilities, and risk sharing in the event of differing site conditions that lead to scope creep, extra costs, and disruption of the project schedule and sequence. Clear, quick, and effective communication, as well as collaboration, would assist in this aspect.

The proactive dispute management (PDM) question is whether potential disputes from site conditions can be prevented, avoided, or doused. The three PDM concepts of D-PAD seem plausible in the site conditions disputes scenario based on the aforementioned outlines. A culture of thorough and effective site investigation and assessments are a dispute prevention measure, as are marshaling out of clear contractual obligations and responsibilities ahead of construction start. Collaborative acceptance and willingness to share risks are dispute-avoidance measures. Clear, quick, and effective communication is a crucial risk-dousing measure. How differing and unforeseen site conditions can significantly modify the original contract and project scope and budget should be envisioned, exhaustively discussed, and agreed upon ahead of signing the contract.

2.4 Change Order or Variation Order by Owner

Change order (or variation order, as it is called in other climes outside North America) refers to any modification of the scope and schedule of work that would add to or subtract from the contract sum. While addenda apply before the contract is signed, any change after that contract that modifies the contract sum is processed as a change order. If the change doesn't modify the contract sum, it is defined and treated as a field order. Both change order and field order can be activated by field directives. Change orders can be anticipated in construction due to unforeseen circumstances like unexpected site conditions, discrepancies, incomplete drawing details and specifications, or a request for additional work or a change in the works.

A change order proceeds thus:

1) *Request for information (RFI) from owner and representatives*: The contractor requests information on, for instance, unexpected site conditions, discrepancies, incomplete drawing details and specifications, or a request for additional work or a change in the work.
2) *Owner and representatives review RFI*: The owner and representatives review the RFI, discuss, and acknowledge the cost implications, if any, and approve it for the contractor.

3) *Documentation and payment*: The contract sum is adjusted accordingly and documented continuously as other change orders occur. The additional costs from the change orders are accounted for in the contract sum and the contractor is paid.

Change orders are a crucial instrument in the construction process, introducing flexibility in already-signed contracts and allowing for accommodation of changing project circumstances and requirements. However, if not handled properly, they can be potential triggers of disputes due to several factors that include:

1) *Delays*: The back and forth of RFIs, reviews, and approvals can take time that may slow or stop aspects of the project, leading to time extensions, and the owner may misconstrue that the contractor is not managing the delays well enough, as to minimally affect the schedules.
2) *Cost argument dispute*: Even where they have agreed unit prices, the cost argument may still arise between the contractor and the owner if compelling documentation and communication on pricing are absent.
3) *Scope creep*: While the owner is not amused by cost overruns occasioned by change orders, the contractor is less amused by scope creep. In the same way, scope creep is substantial and the additional cost of such a request for time extensions may be viewed by the owner as fleecing by the contractor.
4) *Potential communication breakdown*: Communication can break down between the owner and contractor, especially if construction progress does not proceed at the expected pace.
5) *Project quality*: Project quality may be affected, especially where there are too many RFIs and change orders.

In order to minimize, mitigate, or eliminate disputes arising from change orders, their number needs to be kept to the barest minimum by thoroughly working out every detail and circumstance ahead of signing the contract. The parties to the contract need to have a clear understanding of the change order process and be open to accepting the scope, cost, and schedule change implications.

The PDM question is whether potential disputes from change orders can be prevented, avoided, or doused. Just as applicable to addenda, preventing disputes from change orders appears far more difficult, if not impossible, than avoiding or dousing them. The reason is that the mere fact of change orders being in the process portends the presence of these triggers, latent or otherwise. Avoiding them seems more plausible, as change orders originate principally from owners of the projects and their representatives and can be as varied as they come. One suggestion to avoid disputes from change orders is to thoroughly work out all details and circumstances prior to signing the contract, and if and when they occur, to follow due diligence in compellingly documenting and communicating pricing to the owners and representatives. The potential or embodied DRA should be conducted to determine how they can be avoided, and if disputes insist on flaring up, to douse them swiftly and effectively before they escalate. How their addition may significantly modify the terms and conditions of the original contract and project scope should be communicated effectively to the owners and representatives.

2.5 Scope Creep

Scope creep has been briefly discussed under the "change order" section, to which it is connected and has implications. It refers to the increase in the size and content of the project beyond the original contract and contract sum. It occurs when there is an addition of requirements, changes, and features different from or additional to what the project started with. It can occur at any stage of the delivery process, including advanced stages.

Typically, scope creep occurs as a result of:

1) *Poor project planning and approach*: Poor project planning starts with haphazard feasibility studies and poor programming, leading to missed features and requirements. Hence, design alterations or redesign, unforeseen but necessary accommodations and provisions, as well as owner's requests can occur in the middle of an ongoing construction.
2) *Incremental additions*: Minor additions and modifications may be inconsequential to the scope, but when they continue in incremental proportions, scope creep is inevitable, with the accompanying implications.
3) *Afterthought changes and inclusions*: The owner may have new realizations in the middle of a project, which, if implemented, will constitute a significant scope change. For instance, on a previous project where an elevator inclusion was not considered, the owner demanded that midway, when the slabs had been built. Including the elevator resulted in a redesign and a significantly changed scope of work.
4) *Poor and ambiguous communication*: The owner may not have communicated all the needs, and the designer may not have exhaustively asked the necessary questions. It means they will waffle through the construction process, adding and subtracting work at whim.

If not managed and handled properly, scope creep can be a potent trigger of disputes due to several factors that include:

1) *Contractual issues*: If scope creep has not been clearly defined, it could be a source of dispute. Inclusion of agreed unit prices could be a help to handle the added scope of work, but there's no guarantee.

2) *Misaligned expectations of the parties*: Scope creep can introduce unexpected project elements, and agreement on how to quantify and charge for them, which may be a cause of dispute between the parties.
3) *Cost overruns*: Scope creep requires more resources, and the increased cost can be a cause of dispute between the parties as to who bears the costs and what is reasonable.
4) *Delays*: Scope creep requires extra time with concomitant contract time extensions and liquidated or ascertained damages, either of which can be disputatious between the parties.
5) *Compromised quality*: Project quality may suffer due to scope creep and the quality of material and workmanship probably used to minimize costs and time extensions. In the end, both parties may be dissatisfied, and this can lead to dispute with release of retention and final payment.

In order to minimize, mitigate or eliminate disputes arising from scope creep, limit it to the barest minimum by thorough and comprehensive programing and design documentation. The parties to the contract need to have a clear understanding of the project requirements, and if scope creep occurs, to be open to mutually accepting the increased scope, cost, and schedule change implications.

The PDM question is whether potential disputes from scope creep can be prevented, avoided, or doused. Just as applicable to addenda and change orders, preventing disputes from scope creep appears far more difficult, if not impossible, than avoiding or dousing them. The reason is that the possibility of scope creep in any project cannot be predicted. Avoiding disputes from it seems more plausible as scope creep originates from the owners of the projects and their representatives. One suggestion to avoid disputes from change orders is to ensure thorough programming and comprehensive design prior to signing the contract, and if and when it occurs, to follow due diligence in compellingly documenting and communicating extra costs and schedule extensions to the owners and representatives. The potential or embodied DRA should be conducted to determine how they can be avoided, and if disputes insist on flaring up from scope creep, to douse them swiftly and effectively before they escalate. How their addition may significantly modify the terms and conditions of the original contract and project scope should be communicated effectively to the owners and representatives.

2.6 Project Turnover or Handover Hitches

Project turnover (or handover, as it is called in other climes outside North America) refers to the transfer of a completed project from the contractor back to the owner. When the contract is signed between the contractor and the owner, the former takes possession of the project and turns or hands it over to the latter upon completion. Several steps are taken to effectively hand over a completed project:

1) *Contractor advice to owner*: When, in the opinion of the contractor, the project has reached practical or substantial completion, the next step is to go through handover processes to allow the owner to occupy the project for the purpose for which it was built. Practical or substantial completion means that the project is not yet one hundred percent complete, but it is occupiable, as the remaining work or corrections cannot obstruct its use for the purpose for which it was built.
2) *Walk-through inspections*: Prior to the turnover, the contractor with the owner and representative(s) walk through the project to thoroughly inspect and confirm that it has reached substantial completion, is occupiable, and that it has been built to specifications, codes, and standards. The walk-through looks for defective, unsafe, and incomplete work that must be completed before or after handover.
3) *Punch list*: The punch list is a list of observed defective, unsafe, and incomplete works that must be completed before or after handover, as ascertained from the walk-through. The contractor undertakes to complete those before the project is certified as finally completed.
4) *Systems commissioning*: For a project of any size, it is almost a given that heating, ventilating, and cooling (HVAC) systems as well as electrical, plumbing, and fire suppression systems will be commissioned. Building envelope commissioning is also becoming commonplace nowadays, especially with complex cladding and curtainwall systems. By commissioning, you test and verify the robustness of these systems to ensure that they are properly installed to function properly, effectively, and efficiently.
5) *Manuals and warranty*: Like vehicles and automobiles, building projects are complex systems that deserve a manual to enable occupants and users to understand and fully put them to maximum use. Manuals help in this process and are prepared and handed over to the owner by the contractor. Some contractors are introducing secondment as a supplement to manuals, where they leave behind a staff member or two with the owner organization for a few years to help take them through the understanding and full use of the built and handed-over facility. This practice is in recognition that building projects are very complex systems that require more than a manual to understand and put to full use. The project owner organization or the occupants and users are oriented and trained on the operation and maintenance of the project as necessary.
6) *Warranties*: Warranties as applicable to the project and parts of the project are handed over to the owner by the contractor, warranting that the project and parts will reliably serve the owner effectively for a specified period of

time during which any failures will be fixed by the contractor or component manufacturers.

7) *As-built drawings*: The many changes along the building of a project could mean that the starting drawings may be different from the finishing drawings, and these differences are captured in as-built drawings that start off the maintenance management of the project through its lifecycle. Bulletin and change order drawings are progressively collected and documented by the contractor and handed over to the owner and representative design team, who update the original drawings to as-built and archivable drawings for operational and maintenance management.
8) *Turnover of access and keys, as well as insurance*: The contractor gets the project to a safe condition for unfettered access by the owner, occupants, and users with keys and other access tools and protocols. The insurance of the project becomes the responsibility of the owner at this point.
9) *Release of final retention, final payment, and a waiver of lien*: On final project completion, the owner releases any retained amounts due to the contractor, as well as the final payment on the contractor's completion of the punch list items. If specified in the contract, the contractor may offer a waiver of lien to the owner ahead of receipt of the final payment, guaranteeing that nobody will place a lien on the project or prevent its continued use in any way.

A successful project wrap-up and turnover by the contractor are crucial to the owner's overall satisfaction with the project, ensuring a seamless transition from construction to occupation, and averting disputes that can occur otherwise. Collaboration of the project participants and stakeholders is essential to achieving this success.

Project turnover is a crucial phase of the construction process, marking the end of the project journey. If not handled properly, it has the potential to trigger disputes due to several factors that include:

1) *Defective and incomplete work*: There may be a dispute over the completeness of the work, defective work, and who should take full responsibility for them. This may relate to the quality of work not meeting the expectations of the owner.
2) *Delays*: Wrap-up can be a problem, with many contractors ineffectively turning over the project. Some contractors can speed up the construction delivery process, but when it gets to be time for wrap-up, it may be found wanting. Some reasons for this scenario may be connected with robbing Peter to pay Paul, suggesting that they might have misused the project funds. Some take a lot of time gathering all the documentation for turnover. These delays could be triggers of disputes between the parties.
3) *Turnover documentation and communication*: There can be a lot of documents associated with a complete turnover. Hence, there may be missed

documentation and the necessary communications in this milieu. Such can be a cause of disputes between the parties.

4) *Clarity of contractual obligations at turnover*: Sometimes, the contract may be unclear or ambiguous as to the obligations of the parties at turnover—what is expected of the contractor and the owner during this stage. For instance, when does the builder's insurance terminate on the project so that the owner's policy can kick in? Such can be a cause of disputes between the parties.
5) *Release of retention and final payment*: Some owners tend to make onerous requests as a condition to release retained and final payments to the contractor. For instance, the owner may demand from the contractor a waiver of lien as a condition for the release of the retention and final payment, both of which might not have been included in the contract. There may also be disagreements if incomplete and defective works creep beyond the scope of the contract, about which the contractor can reject responsibility.

In order to minimize, mitigate, or eliminate disputes arising from project turnover, the parties need to be thorough in the contract with regard to turnover, with the expectations and responsibilities clearly spelled out. The applicable documentation should be complete and comprehensive, and if scope creep occurs at this stage, the parties should be open to mutually accepting the increased scope, cost, and change implications, with the release of retention and final payment made promptly.

The PDM question is whether potential disputes from project turnover can be prevented, avoided, or doused. The three PDM concepts of D-PAD seem plausible in project turnover disputes scenarios based on the aforementioned outlines. A culture of thorough and effective communication is a dispute-prevention measure, as is the marshaling out of clear contractual obligations and responsibilities ahead of construction commencement. Collaboration of the parties during turnover and willingness to share risks are dispute-avoidance measures. Clear, quick, and effective documentation, communication, and implementation are crucial risk-dousing measures. The potential or embodied DRA should be conducted to determine how they can be avoided, and if disputes insist on flaring up during turnover, to douse them swiftly and effectively before they escalate.

2.7 Constructive Acceleration, Lofty, and Unrealistic Expectations by Owner

Constructive acceleration by the project owner occurs when the owner causes the contractor to speed up the work to beat, better, or meet the project deadline despite any prior delays beyond the control of the contractor, such as inclement weather, force majeure, or the act of God, including delays from the owner and representatives. Constructive acceleration can come with lofty and unrealistic expectations from the owner, and take the form of:

1) *Owner challenge with bonus and penalty*: The owner can acknowledge delays caused by the representatives, inclement weather, or force majeure, but challenge the contractor to still aspire to be on schedule to deliver the project on time or earlier, with a specified bonus if achieved, or a specified penalty if not.
2) *Indirect owner pressure and change of schedule*: The owner may subtly communicate to the contractor to speed up the project and deliver before the deadline specified in the contract, without any incentive for accelerated delivery. It means delivering the project within a compressed timeframe at variance with the contractual schedule.
3) *Approval delays*: The owner may delay approvals of the contractor's submittals, shop drawings, change orders, and certificates of payment. If the work is delayed, the contractor will be under pressure to still deliver the project according to the contractual deadline without any further compensation for delays caused by the owner.
4) *Changed scope*: The owner may demand a change in the scope of the project without correspondingly changing the schedule or the delivery deadline.

Constructive acceleration can be a contentious issue in any project delivery process because of the sometimes unacknowledged financial and resource pressures it levies on the contractor. The owners should recognize the implications of their whimsical requests on the contractor, as they can put immense pressure on and strain the relationship with the contractors, leading to potential disputes, especially on the demand for fair compensation for constructive acceleration.

Constructive acceleration can be a potential trigger for dispute between the contractor and the owner due to the following:

1) *Extra costs*: Constructive acceleration can come with extra costs for the contractor, which the owner may not completely acknowledge, thus leading to potential disputes between the parties.
2) *Pressure on schedule*: Constructive acceleration can compress or pressurize the project schedule, and the contractor is still expected to deliver on the schedule as wished by the owner, with the accompanying additional resources and demands, such as costs, overtime, more human resources, and equipment. Such can be a cause of disputes between the parties.
3) *Contract variation*: As constructive acceleration can be at variance with the contract in terms of scope, schedule, and contract sum, the contractor won't breach the contract by refusing to carry out the extra work. However, this refusal can cause dispute as to the responsibilities and entitlements of the

parties, and the contract may not have been clear on or made provisions for this scenario.

4) *Communication between the parties*: As constructive acceleration from the owner sometimes comes as subtle pressure and creation of circumstances to force the hands of the contractors, communication can be ambiguous and prone to misinterpretations. Such can deteriorate into disputes between the parties.

In order to minimize, mitigate, or eliminate disputes arising from constructive acceleration, the parties need to be explicit in the contract with regard to handling constructive acceleration if and when it occurs, with the responsibilities and entitlements clearly spelled out. The applicable incentives should be in place, and if contractual schedule shortening occurs as a result, the parties should be open to mutually accepting any cost and damage implications.

The PDM question is whether potential disputes from constructive acceleration can be prevented, avoided, or doused. The three PDM concepts of D-PAD seem plausible in constructive acceleration dispute scenarios based on the aforementioned outlines. A culture of working with explicit contracts and effective communication is a dispute-prevention measure, as are marshaling out of clear contractual obligations, responsibilities, and entitlements ahead of construction commencement. A collaborative approach of the parties to constructive acceleration—and willingness to share the risks and the rewards—are dispute-avoidance measures. Clear, quick, and effective documentation, communication, and implementation are crucial risk-dousing measures. The potential or embodied DRA should be conducted to determine how disputes can be avoided, and if disputes insist on flaring up on the occurrence of constructive acceleration, to douse them swiftly and effectively before they escalate.

2.8 Payment Delays

Payment delays by the project owner can occur due to several reasons:

1) *Cash flow problems*: Depending on the type of financing, an owner dependent on cash receipts from external sources can run into cash flow problems resulting in delayed payments to the contractor, and by extension, to subcontractors and suppliers. The owner may also misuse or misapply the project funds on ventures or projects other than the one at hand. Where the project is financed by the bank, the process of payment may not be prompt, as the bank schedules and undertakes inspections ahead of approving and paying the contractor.
2) *Disagreements*: There can be disagreements over the quantity and quality of work, materials purchased and placed on or stored near the project, the extent

of scope change, change order costing, discrepancies, and ambiguities. Such disagreements may take time to work out, thus leading to delays in payment by the owner.

3) *Documentation and approval time*: While the project is in progress and a change order or change of scope occurs, accompanied by documentation that would need approval, this may take time, thus causing delays in getting to the point of payment by the owner.
4) *External factors*: Owner's payment delays may be the result of external factors, such as a sudden economic downturn, unforeseen circumstances, natural disasters, and regulatory changes. For instance, COVID-19 was unforeseen, and it took all by surprise, and project owners and financiers were not exempt.

Payment delays have serious cashflow consequences on the contractor, subcontractors, sub-subcontractors, and supplies, and such can strain the relationship of the parties and lead to disputes. Where payment delays are unavoidable, they must be handled with transparency, communicated effectively, and addressed swiftly.

Payment delays by the owner are a common cause of disputes between the parties. When the owner fails to make timely payments to the contractor, a chain effect is created where subcontractors and suppliers may not be paid as well. Payment delays by the owner cause disputes due to the following reasons:

1) *Cash flow problems and financial strain on the contractor*: Payment delays by the owner come with chain effects on the contractor, subcontractors, and suppliers. Ethically, the contractor cannot source funds from other projects to pay for labor, material, and equipment used on the project at hand. The same applies to subcontractors and suppliers. Quibbling over delayed payments can strain the relationships between the parties and invite contested disputes if not expeditiously addressed.
2) *Disruption of workflow*: Cash flow can disrupt workflow, and this situation affects the interests of both parties and can lead to disputes sooner or later. The contractor may halt the work or slow down operations, which can lead to the need for disputatious time extensions and damages.
3) *Work quality*: Work quality may suffer as a result of delayed payments, leading the contractor to cut corners to keep the project going. Poor work quality is a certain trigger of disputes.
4) *Breach of contract*: Payment delays can force the contractor to breach the contract by halting the work or delaying operations to the point of tampering with the contractual timeline to deliver the finished project.
5) *Late payment interest*: Some contracts may specify late fees for delayed or late payments, and sometimes this can breed argument, especially when the terms are not well-spelled out with the responsibilities and entitlements clearly defined. For public projects with the statutory Prompt Payment Act,

> the argument may be less contentious, but that is not the case with private projects. For instance, as stated by Clough et al. (2015), the Prompt Payment Act provided that the government is required to make payments to prime contractors within fourteen days of receiving a progress payment application or pay the contractor interest at the Prompt Payment Act interest rate established by the Contract Disputes Act. The prime contractor, in turn, is required to pay its subcontractors within seven days of receiving payment from the government or pay interest. The prime contractor is allowed to withhold payment for cause, such as deficient performance. The prime contractor must also certify to the government that all payments from the government have been earned, and that subcontractors and suppliers have been paid out of previous payments.

In order to minimize, mitigate, or eliminate disputes arising from payment delays by the owner, the parties need to satisfy themselves with the certainty of funding availability and its prompt disbursement when due. Interest or penalty on a late payment should be clearly spelled out, whether in the contract or in a memorandum of understanding (MOU), with the responsibilities and entitlements clearly delineated. Liquidated and ascertained damages should be defined in the contract, and if a project time extension occurs as a result of payment delays, the parties should be open to mutually accepting any cost and damages implications that must be promptly communicated and swiftly consented to.

The PDM question is whether potential disputes from payment delays by the owner can be prevented, avoided, or doused. Only two PDM concepts of D-PAD, namely dispute prevention and dispute dousing, seem plausible in a payment delay dispute scenario based on the aforementioned outlines. A culture of working with comprehensive contracts, including details such as penalty and bonus clauses, is a dispute prevention measure, as is marshaling out of clear contractual obligations, responsibilities, and entitlements in delayed payments ahead of construction commencement. Clear, quick, and effective documentation, communication, and implementation are crucial risk-dousing measures. The potential or embodied DRA should be conducted to determine how risks can be avoided, and if disputes insist on flaring up on the occurrence of delayed payments, to douse them swiftly and effectively before they escalate.

References and Suggested Further Reading

Clough, R.H., Sears, G.A., Sears, S.K. et al. (2015). *Construction Contracting – A Practical Guide to Company Management*, 8e. Hoboken, NJ: John Wiley & Sons, Inc.

Chapter Review Questions

1) Addenda issued late in the bidding process can lead to disputes due to communication timing issues. True/False?
2) Scope creep always occurs during the early stages of a construction project. True/False?
3) Constructive acceleration refers to the owner's demand for faster project delivery despite delays beyond the contractor's control. True/False?
4) Payment delays by the owner have no impact on contractor cash flow or project progress. True/False?
5) Project turnover is a straightforward process that rarely causes disputes. True/False?
6) Addenda can sometimes conflict with other provisions of the contract, leading to disputes. True/False?
7) Which of the following is NOT a common owner-related dispute flashpoint?
   a) Addenda
   b) Site conditions
   c) Contractor's financial capacity
   d) Scope creep
8) What is a key strategy to avoid disputes from addenda?
   a) Ignoring them during bidding
   b) Exhaustive clarification and documentation
   c) Relying on verbal agreements
   d) Delaying contract signing
9) Which of the following is a potential consequence of scope creep?
   a) Reduced project cost
   b) Improved quality
   c) Project delays and cost overruns
   d) Elimination of change orders
10) Which PDM concept is most applicable when disputes begin to emerge and need to be quickly neutralized?
    a) Dispute prevention
    b) Dispute avoidance
    c) Dispute dousing
    d) Dispute resolution
11) Which of the following is a recommended approach to mitigate disputes during project turnover?
    a) Avoid inspections
    b) Delay final payment
    c) Ensure thorough documentation and clear communication
    d) Ignore punch list items

12) ________ refers to the increase in project scope beyond the original contract.
13) A ________ is issued to modify the contract after signing, often due to unforeseen conditions or owner requests.
14) ________ acceleration occurs when the owner indirectly pressures the contractor to speed up work despite delays.
15) During project turnover, a ________ is created to list incomplete or defective work.
16) The PDM concept of ________ involves taking steps to prevent disputes from arising.
17) Which of the following is a site condition that can lead to disputes?
    a) Clear weather
    b) Flat terrain
    c) Unmarked underground utilities
    d) Standard soil type
18) Which of the following is NOT a recommended strategy for handling payment delays?
    a) Clear contract clauses
    b) Open communication
    c) Ignoring the issue
    d) Risk analysis
19) Project turnover disputes can arise due to unclear ________ obligations.
    a) Financial
    b) Contractual
    c) Legal
    d) Environmental
20) Which of the following is a key benefit of applying PDM concepts to owner-related disputes?
    a) Increased litigation
    b) Reduced collaboration
    c) Hitch-free project outcomes
    d) Higher project costs

Exercises

Exercise 2.1 Addenda Risk Assessment Simulation

Objective:
Understand how addenda issued during the bidding phase can lead to disputes, and apply PDM strategies to mitigate risks.

Instructions:

1) Choose a real or hypothetical construction project with multiple addenda issued during the bidding phase.
2) Review each addendum and identify:
   - Changes to scope
   - Potential ambiguities
   - Timing of issuance
3) Conduct a mock dispute-risk analysis (DRA) using PDM principles:
   - What disputes could arise from each addendum?
   - Which D-PAD strategy (prevention, avoidance, dousing) would be most effective?
4) Present your findings in a short report or group discussion.

Outcome:
Gain practical insight into how early documentation changes can trigger disputes and how proactive strategies can mitigate them.

Exercise 2.2 Site Conditions Conflict Resolution Roleplay

Objective:
Explore how unforeseen site conditions can lead to disputes, and practice applying PDM concepts to resolve them.

Instructions:

1) Form a small group and assign roles: Owner, Contractor, Geotechnical Engineer, and Project Manager.
2) Use a real or fictional scenario where unexpected underground conditions (e.g., contaminated soil or unmarked utilities) are discovered after project start.
3) Roleplay a meeting where the team must:
   - Identify the dispute flashpoint
   - Discuss responsibilities and liabilities
   - Apply D-PAD strategies to prevent escalation
4) Reflect on how communication and early risk analysis could have changed the outcome.

Outcome:
Develop skills in collaborative problem-solving and dispute dousing in high-stakes construction scenarios.

Discussions

1) How can project owners proactively manage change orders to minimize disputes with contractors? Discuss the role of clear communication, thorough

documentation, and early planning. Explore how PDM concepts like dispute avoidance and dispute dousing can be applied when changes are inevitable.

2) In what ways do unrealistic expectations from project owners contribute to disputes, and how can contractors respond without escalating conflict? Consider examples such as construction acceleration or scope creep. Debate strategies for maintaining professionalism and applying PDM principles to preserve relationships and project outcomes.

3

Contractor-Related Dispute Flashpoints

Readers' Learning Objectives and Outcomes

1) *Analyze the various contractor-related flashpoints and triggers of conflicts and disputes in the industry.*
2) *Conceptualize plausible prevention, avoidance, or dousing approaches that may be applicable to these conflicts' and disputes' flashpoints and triggers.*
3) *Match and apply prevention, avoidance, or dousing strategies effective or suitable to each conflict's and dispute's flashpoints and triggers.*

3.1 Introduction

The major contribution of the contractor to the construction project process is the ability to marshal out and allocate the resources of funds, materials, labor, and equipment to accomplish it effectively and efficiently in terms of delivery time, budget, and quality. Depending on whom the contractor conducts its business with, it can be a potent source of disputes. Contractor dispute flashpoints and triggers can come from several scenarios, such as:

- financial capability
- progress delays
- liquidated and ascertained damages
- misuse of payment
- nonpayment of subcontractors and suppliers
- technical capability
- labor and equipment productivity variations
- inclusion of alternates
- underbidding and accuracy of estimates
- unrealistic scheduling, project resourcing, and balancing
- use of union and nonunion labor
- work quality
- workers' fidelity
- accidents and safety
- experience modification rating
- mechanics lien
- bonds
- insurance

3.2 Financial Capacity

The financial capacity of a construction contractor refers to the ability to manage organizational resources to sustain the business of successfully delivering projects. A contractor is often measured by surety and insurance companies on the three-C alliteration of character, capacity, and capital. The contractor's financial capacity consists of several aspects:

1) *Access to capital*: A financially capable contractor should have access to capital, whether through self-savings, shares, banks, insurance companies, trusts, or government agencies. With these funds, it can pay for labor and materials, hire equipment, and offset associated business overheads. It must maintain a

clean credit history and a high rating to facilitate borrowing or obtaining lines of credit as needed and at favorable rates.

2) *Management of cash flow*: As the contractor may be handling multiple projects and is paid at interims on completion of portions of the project, it must ensure enough internal liquidity to cover day-to-day or week-to-week expenditures ahead of the interim payments. It should not resort to the proverbial robbing of Peter to pay Paul to cover recurrent daily or weekly expenditures on one project from another.
3) *Contractor profitability*: Profitability is crucial to the contractor's financial capacity. Profitability enables the generation of adequate capital to offset expenditures, reinvest to grow the business, and provide returns to investors.
4) *Financial planning and management*: This aspect involves effective distribution and management of the budget to ensure project completion within the budget. Project expenditures and revenues must be well-documented, tracked, and controlled with financial accounting and planning best practices that include financial statements and balance sheets.
5) *Contingency fund*: The contractor must set aside extra funds to weather sudden and unexpected market and economic downturns that can jeopardize effective project delivery.

Essentially, the financial capacity of the contractor is critical to the success of its construction project deliveries and business sustenance. By applying prudence and best financial planning and management practices, the contractor can enhance its reputation and ensure its competitiveness.

The financial incapacity of a construction contractor can be a common cause of disputes between the parties. When the contractor fails to make timely payments to subcontractors and suppliers, a chain effect can be created that would affect the progress of the project, and this can cause disputes due to the following reasons:

1) *Delayed payments*: Due to the financial incapacity of the contractor, payments for the subcontractors, suppliers, and workers may be delayed, and this can cause disputes.
2) *Cutting of corners*: The financial incapacity of the contractor can lead to cutting corners in terms of materials, labor, and equipment input, and this can affect quality and workmanship, which can bring about disputes.
3) *Stalled project progress*: The financial incapacity of the contractor, leading to delayed payments, also translates to delayed projects. Untimely completion of the project leads to potential disputes with the owner.
4) *Breach of contract*: The financial incapacity of the contractor can bring about a chain effect that extends from payment and project delays to cutting corners and to a breach of the contract. It is a breach of contract to not advance the

project delivery in a timely manner so as to exceed the agreed completion date. A breach of contract attracts potential disputes.

5) *Additive constructive changes*: When there is a change order that changes the scope and cost of the project, the financial incapacity of the contractor will make it more difficult for it to accommodate the changes and the extra costs. Hence, disputes are more likely to arise as a result of such changes.

Essentially, the financial incapacity of the contractor is a critical factor that can affect various aspects of the project and lead to potential disputes if extra care is not taken to proactively manage and prevent or avoid them.

Please refer to Table 14.1 for the recommended PDM's D-PAD concept applicable in this regard.

3.3 Progress Delays

Payment delays can significantly affect the progress of the project, and they often occur due to the following reasons, a couple of which have been mentioned in the earlier sections:

1) *Cash flow problems and financial incapacity of parties:* There can be cash flow issues with the owner, the contractor, and the subcontractors, due to financial incapacity. When a paucity of funds to promptly pay suppliers and workers, delays can cascade down the chain.
2) *Scope creep and change order*: There can be disagreements over payment for a change order or increased scope beyond the contract stipulations. Such disagreements can lead to payment being withheld until a consensus agreement is reached.
3) *Poor quality and workmanship*: Defective and incomplete work can lead to payment delays until corrected, reworked, or completed.
4) *Approval and documentation time*: Payment may be held up due to the bureaucratic processes of approving invoices, certificates, changes, submittals, etc., as well as the processing and disbursement of payments. Paperwork may be erroneous or incomplete, and rectifying it can take extra time and cause payment delays.
5) *Project disruptions*: Project disruptions or delays of any kind (man-made or by act of God) lead to payment delays. Unforeseen circumstances, industrial action of labor, material shortages, sudden economic downturn, inclement weather, etc., can all bring about project disruptions and subsequently delay payments.
6) *Mechanic's lien and claims*: When a claim is against the project by way of a lien by the contractor or subcontractor, it delays the project progress, and by

implication, delays payments to the parties until the claims are addressed and the lien is lifted.

In summary, prompt payments are essential to the smooth operation and progress of any project. By prioritizing clear payment terms and procedures in the contract, the parties can mitigate the risk or occurrence of delays and disputes over them.

Payment delays for whatever reason can be a common cause of disputes between the parties. When the contractor fails to make timely payments to subcontractors and suppliers, or to pay for materials and equipment, a chain effect can be created that would affect the progress and workmanship of the project, and it can cause disputes due to the following reasons:

1) *Financial strain*: Delayed payment from one party to another can put financial strain on the party not timely paid, and this can escalate into disputes. Cash flow is disrupted, thus making it difficult to meet the unpaid party's financial obligations.
2) *Breach of contract*: The payment terms and procedures are often spelled out in the contract. Hence, a delayed payment would be viewed as a breach of contract, and that can trigger disputes over contractual obligations, rights, and remedies.
3) *Stalled project progress*: As delayed payments can disrupt the progress of the project, disputes can arise over the contractual timelines and schedules, and over which party to attribute the responsibility for resultant delays or halting of the works.
4) *Lowering of morale, goodwill, and trust*: Where payments are delayed, morale, goodwill, and trust between the parties can be eroded and relationships strained, thus setting the stage for frictions, conflicts, and disputes.
5) *Mechanic's lien and its effects*: Where a lien is placed on the project in progress by an unpaid party, this can escalate things into a dispute that may be mediated, arbitrated, or litigated.

Overall, delayed payment has the potential to escalate into disputes that can hurt the smooth running and progress of the project, strain the relationships between the parties, and negatively affect the success of the project. It is crucial for the parties to address payment issues swiftly and professionally to avoid disputes and maintain productive working relationships on the project. Clear communication and adherence to interim and timely payment obligations in the contract can help to mitigate or avoid payment delays and subsequent disputes from them.

Please refer to Table 14.1 for the recommended PDM's D-PAD concept applicable in this regard.

3.4 Liquidated and Ascertained Damages

Liquidated and ascertained damages are often stated in the contracts to incentivize the timely completion of the project. The reason is that the project is meant for a business that may be rendering a service and earning money, so that each day it is delayed from being put to use for the service or business means a loss of earnings. Their respective definitions are as follows:

1) *Liquidated damages*: They refer to pre-estimated and pre-agreed compensations and rates for a breach of contract, mostly to do with the project completion timeline. Liquidated damages deal with projected estimates of costs and liabilities, as well as rates agreed between the parties ahead of time for when the project exceeds its completion timeline. This means that daily or weekly damages levied may exceed or be below the actual liability. That's where the give and take or differing views of the parties can differ and conflict.
2) *Ascertained damages*: They are concerned with actual costs and entitlements resulting from a breach of contract with regard to the project completion timeline. Ascertained damages are concerned with actual costs and liabilities resulting from the project exceeding its completion timeline.

Liquidated and ascertained damages, if properly applied, facilitate the timely completion of the project and are win-win for all parties. They can be put to service or business without delay, and this can begin to generate income. The impact of liquidated and ascertained damages is most notable under the Eichleay formula, especially with large contractors that have greater overhead costs. This formula is a method of calculating delay damages from a construction company's office overhead costs, such as payroll, benefits, taxes, utilities, and insurance.

Liquidated and ascertained damages are meant to provide clarity and deter breach of contract timelines. However, if misapplied, they can be triggers of disputes between the parties, and here are the reasons why:

1) *Liquidated and ascertained damages as punitive measures*: Liquidated and ascertained damages are best applied as incentivization measures for the timely completion of the project. However, when they are used with punitive intent, they can set the stage for adversarial and disputatious relationships between the parties.
2) *Excessive and unreasonable liquidated and ascertained damages*: Where liquidated and ascertained damages are deemed as excessive and unreasonable by the contract-breaching party, they can be cause for disputes. Determining the reasonable amount can be contentious, which is why some contracts wisely provide not just liquidated damages, but both liquidated and ascertained damages, so that either can be applied where deemed more appropriate or reasonable.

3) *Apportioning responsibility for delays*: As monetary liability is involved in liquidated and ascertained damages, disputes may arise as to the validity of this clause, who actually caused the delays, and to what extent. The contractor can argue that the causes of delays were beyond the contractor's control, such as late material deliveries, and design errors and omissions that are attributable to the side of the privity of the contract. What constitutes delays leading to completion time extension and damage claims could be points of contention between the parties.
4) *Mitigation of damages*: The party to pay damages may dispute that the contract's nonbreaching party failed to take steps to mitigate and minimize the damages. For instance, in a situation of force majeure or act of God involving a hurricane, the contractor failing to take steps to mitigate the impacts and subsequent delays may find it hard to claim full damages from shrewd or difficult owners. This can lead to disputes between the parties.
5) *Excusable delays and damages*: Delays occasioned by act of God, changes, and scope creep attributed to the owner, or unanticipated regulatory changes may not be subject to damage claims. Arguments may arise as to what falls under excusable delays and damages, and this can be a trigger for disputes between the parties.

Overall, a properly applied liquidated and ascertained damages clause provides a degree of certainty and predictability in construction contracts, and it helps to mitigate and prevent disputes caused by untimely completion and delivery of the projects.

Please refer to Table 14.1 for the recommended PDM's D-PAD concept applicable in this regard.

3.5 Progress or Interim Payments and Their Misuse

Progress or interim payments refer to either regular intervals of payment by the owner to the contractor monthly or even fortnightly as the project progresses, or payment based on agreed milestones from the start to the completion of the project. The contractor then pays the subcontractors and suppliers. Progress payments proceed in the following manner:

1) *Contractual terms of payment*: The owner and the contractor work with a contract that outlines the scope of the project, its time schedule, and the payment terms. The contract can specify the schedule of values and progress payment conditions.
2) *Payment schedule and measurement of work*: Where payments are to be staged and disbursed at specified interims, they are often preceded by the

measurement of the amount of accomplished work and materials delivered within or near the project to ensure that the payments are equivalent to the work accomplished and the materials delivered. Payments are often tied to percentages of project completion or specified project milestones, stages, or phases. For fixed-price contracts, the contractor is supposed to submit applications for progress payments at least 10 days before the established payment interims.

3) *Documentary evidence of claimed completion*: The contractor must present documentation and evidence of the claimed completion for which payment is demanded or due. Examples are invoices, evidence of payment for all materials from suppliers, and labor from subcontractors, besides the physical and measurable completed works.
4) *Approval of payment claim and issue of certificate of payment*: When the owner or owner's representative confirms that the project has been accomplished as per the percentage, milestone, or stage of progress claimed, the contractor's payment request will be approved and a certificate of payment issued. Disbursement has to happen as soon as possible, and should not exceed seven days, beyond which, if agreed, interest may begin to accrue on the unpaid payment until paid in full. An agreed percentage of the payment to the contractor, typically 10%, may be retained by the owner as a way to incentivize the progress of the project, and may be tapped if the owner has to pay for aspects of the project that are supposed to be the responsibility of the contractor. In other words, for every payment made to the contractor, the owner withholds 10% of it. For public projects, payment to the contractor by the owner must be made within 14 days of receipt of the payment application, as specified in the Prompt Payment Act Amendments of 1988. Failing to pay within this time frame will attract interest payments to the contractor at the prevailing Prompt Payment Act interest rate established under the Contract Disputes Act, which may vary from 2% to 3% daily on the amount owed until paid in full. In turn, the contractor is expected to pay subcontractors and suppliers within seven days of receipt of progress payment from the government, failing which it must also pay daily interest at the same prevailing rate until fully paid.

Progress payments help to ensure the steady and consistent continuation of the project from start to finish by providing the necessary cash flow to the contractor to discharge all project-related financial obligations—payment for labor, materials, equipment, overheads, etc. The contractor is paid for the work accomplished rather than waiting until project completion, and the owner visually sees evidence and justification of each and every payment disbursement.

However, many contractors are notorious for cash flow issues, and the imprudent ones can misuse their progress payments to offset financial obligations other than those associated with the projects for which they have been paid—the proverbial robbing Peter to pay Paul scenario. Progress payments can be triggers for disputes between the parties, and here are some of the reasons:

1) *Delayed payments*: From application for payment to approval and disbursement, delays can happen that can strain cash flow down the chain. The contractor cannot pay the subcontractors and the suppliers when they are not paid. In fact, some contracts bear clauses that say "Pay when paid," suggesting that the subcontractors and the suppliers can only be paid if and when the contractor has been paid. Delays in payment can trigger disputes between the parties.
2) *Disagreement over the amount of work completed*: Where the percentages of work completed are established by visual assessments or other approximating appraisal, the actual amount and quality of work completed may be open to debate between the parties. In fact, visual assessments of the percentage of completed work have exposed architects and engineers to suits and claims of negligence by the owners. Quantity surveyors and cost engineers may be better equipped than architects and engineers to more accurately measure the quantity and percentages of work accomplished to ensure that overpayment doesn't occur. Disputes can arise between the parties where there are disagreements over the amount of work accomplished or where there might have been overpayments.
3) *Claimed completed work quality*: There may be disagreements over completed work quality or defective work for which a progress payment is claimed. The owner may withhold or delay a progress payment if not satisfied with the work quality for which the payment claim is made, and the contractor may dispute the owner's assessment of the work quality and withholding of payment.
4) *Scope creep and change order disruption of payment*: Scope creep and change of the work beyond the contract can disrupt the schedule of interim payments. Disputes may arise over the true cost of the increased scope or changes, and the owner may refuse to approve commensurate additional payment.
5) *Cash flow issues with the contractor*: The contractor may be misusing progress payments to offset financial obligations not associated with the project at hand. This will have a chain effect down the line in terms of paying the subcontractors and suppliers, who can place a lien on the project and stall its progress to timely completion. Such a scenario is a potent trigger of disputes between the parties.

Essentially, to mitigate and avoid disputes from progress payments and their misuse, it is crucial for the parties to strictly adhere to the letter of the contract on prompt interim payments, including the release of retained payments when due. They should communicate effectively and maintain accurate documents of the work and the progress attained and claimed for, and swiftly address or douse any issues and discrepancies if and when they arise.

Please refer to Table 14.1 for the recommended PDM's D-PAD concept applicable in this regard.

3.6 Character and Personality of Contractor

The character and personality of the contractor are crucial intangible traits and characteristics that determine how successful the contractor and the projects executed can be. These traits and characteristics can vary from one contractor to another. For instance:

1) *Ethics and integrity*: The contractor's adherence to codes of ethics and integrity in the execution and delivery of projects determines its success. An ethical contractor will operate with integrity, which translates to honesty and transparency that promote camaraderie with the owner and other project participants. Such is apparent in the contractor's trustworthiness in all project dealings, from strict compliance with the terms of the contract to financial matters.
2) *Reliability*: The reliability of the contractors can vary. Some are more reliable than others in staying on the job and delivering on time, to budget, and to quality.
3) *Charisma*: Charismatic contractors are effective in leading the construction process, appropriately delegating authority, responsibility, and tasks, and getting others to deliver timely results and quality. They motivate and inspire their workers to offer their best in making the project a collective success.
4) *Adaptability and flexibility*: This attribute is essential in contractors, especially in situations of change and disruptions of the project. They are to adapt to the changed work or scope, and be flexible with the measures and actions to overcome disruptions militating against timely project delivery.
5) *Communication skills*: Contractors with good communication skills enhance the success of project delivery with less hassles. Clear and effective communication is essential on projects, and the contractor needs to clearly communicate both simple and complex information to the owner, subcontractors, and suppliers. The contractor is able to listen actively, negotiate, and navigate effectively to ensure that all participants and stakeholders are on the same page.
6) *Problem-solving skill*: This attribute can vary among contractors and can determine whether a project will be successful or not. Projects inevitably encounter obstacles and challenges, such as technical, logistical, and regulatory issues, and the contractor equipped with ample problem-solving skills is able to identify the issues early to develop solutions and effectively and timely implement them to keep the project on track.
7) *Attention to detail*: This characteristic can vary among contractors, and meticulous attention to detail can make a difference in how issues of quality and safety are handled. Even slight oversights and laxity could lead to costly mistakes, delays, and breaches of safety and quality.

8) *Resilience*: As construction is a highly challenging and demanding industry with tight deadlines, demanding owners, and uncertainties, resilience is an essential attribute for contractors to handle and navigate through the adversities and pressures to timely deliver the project successfully.
9) *Financial management*: The contractor should be able to negotiate the contract, manage the budget, and procure materials cost-effectively to ensure the financial viability of the project. Contractors with prudent financial acumen and know-how to manage project finance with sound accounting systems and understand the need not to divert the project finances to offset other unrelated financial obligations.
10) *Customer satisfaction*: A contractor who is customer-focused understands the necessity to satisfy the owner as a customer for the success of the project at hand and future projects. Hence, such a contractor listens and delivers the needs of the owner, communicates effectively and timely, and strives to meet and exceed expectations to build lasting relationships and possibly earn repeat business and recommendations.

While experience and technical skills are important, the intangible character and personality traits of contractors form a bedrock of their success and the success of their projects. These set exceptional contractors apart from their counterparts, and project owners should seek them out if they desire successful and less-disputatious projects.

However, many contractors come with apparent character and personality flaws that can lead their projects to proceed without some certainty of successful outcomes. Here are some reasons why character and personality flaws of the contractor can be the root causes of disputes:

1) *Poor communication*: The contractor that lacks effective communication skills will convey both simple and complex project information poorly, leading to misunderstandings, delays, conflicts, and disputes. Where a contractor fails to clearly and timely communicate project requirements, timelines, and changes, this can bring about disputes due to associated costly consequences, such as confusion and delays.
2) *Unreliability*: Contractors who often fail to deliver timely and quality projects risk frustrating the owner, subcontractors, and suppliers, and this can escalate into disputes over delays and quality issues.
3) *Questionable integrity*: Contractors lacking integrity can resort to all sorts of untrustworthy practices, like cutting corners, installing substandard materials, and using unskilled labor, thus delivering subpar projects. Such behavior can lead to disputes when the owner uncovers discrepancies in what was delivered versus what was agreed and guaranteed in the contract.

4) *Poor leadership skills*: Contractors with poor leadership skills are ill-equipped with the charisma and capability to effectively manage crews and teams, leading to ineffectiveness, low morale, and possible conflicts and/or disputes. Amid a lack of direction and guidance, the crew and teams will be rudderless, and this contributes to errors, reworks, accidents, and disputes among workers.
5) *Rigidity*: When a contract is inflexible and unadaptable in matters of change or accommodating and meeting the owner's evolving needs as the project progresses, such a stance can shatter camaraderie between them and kickstart an adversarial relationship that would inevitably lead to one dispute or another.
6) *Negligent contractor*: Contractors who are negligent of their responsibilities, such as safety protocols, regulatory compliance, and quality compliance, can create liabilities for themselves and other project participants. Disputes can arise out of property damage, accidents, and other costly issues.
7) *Financial mismanagement*: Contractors who are prone to mismanaging projects invite disputes with the owner, subcontractors, and suppliers. Budget overrun, nonpayment of subcontractors and suppliers, and improper invoicing can all result from financial mismanagement of the contract, and all of this leads to disputes.
8) *Customer satisfaction disregard*: Contractors who do not prioritize owners' satisfaction over their profits run the risk of embroiling themselves in disputes over project quality and the duty of care expressed or implied in the contract.

Overall, while contractors' experience and technical skills are important in the project delivery process, their character and personality are essential in guaranteeing their success and that of their projects. Their character and personality, if flawed, can get in the way and bring about avoidable conflicts and disputes. The sound character and personality of the contractor are crucial in building strong and lasting relationships, fostering open and effective communication, and prioritizing ethics and professionalism to prevent, avoid, or douse conflicts and disputes, guaranteeing successful project outcomes and customer satisfaction.

Please refer to Table 14.1 for the recommended PDM's D-PAD concept applicable in this regard.

3.7 Technical Capability and Experience

The technical capability and experience of a construction contractor comprise range of skills, knowledge, and resources acquired over time that enable the successful accomplishment of project delivery. They include:

1) *Technical expertise*: This refers to the contractor's proficiency in the necessary techniques and methods of project construction, handling of materials,

equipment, and human resources relevant to the type of projects undertaken. They should be able to competently read, interpret, and translate architectural and engineering blueprints and specifications into constructed projects.

2) *Project and resource management*: The contractor's skills in this aspect are crucial to coordinating human, material, and equipment resources, optimizing available finances to successfully deliver projects on time, within the budget, and to quality. Involved here are estimating, budgeting, planning, scheduling, resource allocation, tracking, cash flow management, and coordination of the trades and suppliers.
3) *Familiarity with prevailing regulations and compliance*: The contractor should be current in their understanding and interpretation of the building codes, zoning ordinances, safety regulations, and other statutory or legal requirements of projects. Noncompliance may lead to delays, fines, and unplanned work stoppages.
4) *Quality assurance and control*: It is necessary for the contractor to have some structures of quality assurance (QA) and quality control (QC), or total quality management (TQM), with regard to their workmanship and the whole project from start to finish. There should be a process in place for testing materials, inspecting, monitoring, and addressing any observed deficiencies as the project continues.
5) *Safety practices*: A technically capable and experienced contractor must have in place abiding safety practices to safeguard workers from accidents and injuries of various types and to save themselves from the cost of workers' compensation headaches. They should provide appropriate personal protective equipment (PPE) and strictly enforce safety protocols at the site. Contractors with poor safety records reflected in a high experience modification rating (mod rate) will not succeed or survive in the business of project construction.
6) *Problem-solving skills*: Due to the nature of construction with many often-unexpected challenges, the contractor should be adept at identifying these challenges, figuring out solutions, and quickly deploying them to keep the project on track.
7) *Effective communication skills*: The contractor should be effective in clearly and promptly communicating with all stakeholders (architects, engineers, owners, subcontractors, suppliers, regulators, etc.) to convey project requirements, timeline, updates, and issues, and to ensure that everyone is on the same page.
8) *Business reputation*: This refers to the contractor's track record and reputation for successfully delivering quality projects on time and within the budget.

In summary, the technical capability and experience of a construction contractor are demonstrated through their proficiency at managing all aspects of project delivery effectively, efficiently, and safely from start to finish.

Where the technical capability and experience of a construction contractor are in question, they can be causes of disputes, and here are the reasons why that is so:

1) *Poor workmanship*: Poor workmanship comes with corner-cutting in the delivery of substandard and subpar projects to the dissatisfaction of the owner, the customer. Such can lead to costly defective work and rework. Disputes can arise as the owner observes that the workmanship and/or materials fall short of the expected standards as expressly agreed or implied in the contract. Project delays and disagreements over payment are inevitable under these circumstances.
2) *Scope creep*: Changes in the project that increase its scope can further stretch and expose the contractor's technical capabilities and experience. Inability to manage increased scope can lead to disputes over additional costs, project delays, and the responsibility to deal with scope variations.
3) *Handling discrepancies and breach of contract*: If the contractor is deficient in reading, interpreting, and translating blueprints into built projects, and comprehending and implementing the contract, handling of ambiguities and discrepancies can be difficult, and this can lead to disputes if interpretations lead to poor project quality, calling for costly corrections of wrongly executed work and rework.
4) *Safety concerns*: There may be safety issues where there is a paucity in the technical capability and experience of the construction contractor in deploying good safety practices and enforcing the protocols. Accidents and injuries resulting from such negligence and laxity can slow or halt the project altogether and lead to costly claims, negotiations, settlements, and even litigation. This can be a cause of dispute between the contractor and the owner or third parties.
5) *Communication breakdown*: A contractor who does not have the technical capability and experience may not communicate appropriately and timely, leading to communication breakdown associated with misunderstanding, delays, escalation of tension, erosion of trust, and ultimately disputes.
6) *Financial argument*: Where technical capability and experience are lacking in a construction contractor, financial arguments can arise over payment, cost overruns, and project finance management. Such arguments can strain the parties' relationship and deteriorate to disputes.

Overall, the technical capability and experience of the construction contractor are essential to the successful execution and delivery of the project, and where they are lacking, disputes can arise due to misapplications of methods and techniques, budget and financial management, safety protocols, and miscommunication. In order to prevent such disputes, you must ensure that you work with technically capable and experienced construction contractors with a track record and reputation to successfully deliver results.

Please refer to Table 14.1 for the recommended PDM's D-PAD concept applicable in this regard.

3.8 Labor Productivity Fluctuations

Labor productivity refers to direct labor production rates measured as man-hours or crew-hours taken to accomplish a unit of work, or the number of work units accomplished per workday by a worker or a crew, whichever is applicable. Labor productivity can fluctuate due to the following reasons:

1) *Skill and experience of workers*: Availability of skilled and experienced labor affects productivity fluctuations. Where there is a shortage of skilled and experienced labor, productivity may decrease due to less-skilled and inexperienced workers taking longer to accomplish work and making errors that call for corrections and rework.
2) *Project type*: Labor productivity can fluctuate due to project complexity and scope. The more complex the project, the more chances of decreased productivity from time taken to overcome the complexity, errors from complexity, and time to correct or rework them. A maxim goes, "Twisted buildings torture their builders." This torture refers to the time and effort taken to overcome the complexity, novelty, and scope of the project. Projects with intricate designs and unique features tend to demand more requests for information and changes, and would take more time and resources to complete, thus affecting the productivity level.
3) *New methods and technology*: New technologies and methods can boost productivity when fully learned, established, and used. However, their learning curve may initially slow productivity; so, the productivity level depends on where you are on that curve. An example is building information modeling and clash-detection technologies. It took a while for workers to embrace them, but they are now relatively commonplace, assisting in smarter work and improved productivity.
4) *Material availability*: When material availability is out of sync with the available labor, productivity fluctuation is certain to occur. Where there are delays in the delivery of construction materials or poor-quality materials are delivered, labor productivity can decrease.
5) *Project planning and scheduling*: Project planning and scheduling are essential in optimizing labor productivity in construction. Poor planning, coordination, and sequencing of the trades can result in inefficiencies and decreased labor productivity. For instance, when concrete slabs are cast on site, they will take time to cure and set, and if proper planning and scheduling are absent, it may

result in delays for subsequent trades, leading to downtime for them and slowed productivity.

6) *Inspectors and regulatory compliance*: Inspections, compliance with codes and standards, as well as safety regulations and environmental standards, can affect labor productivity. For instance, if parts and stages of the project have to be inspected by regulatory inspectors before proceeding, work may have to halt to wait for and work with the schedule of the inspectors. Many regulatory compliances may require time and resources to comply, thus slowing down the progress of the work.
7) *Location and weather conditions*: Location plays two parts in labor productivity fluctuations. Labor productivity can vary by region and between countries. For instance, labor productivity for framed construction may be faster in certain southern U.S. locations than in the Midwest or the Northwest. Construction labor productivity between China and the United States can vary significantly due to many reasons, including strict regulatory and safety enforcement. Construction labor productivity for skyscraper projects in Hong Kong is far higher than that of Sydney due to the former's greater experience in building tall structures (Picken and Ilozor 2003, 2014). Locations can vary in their weather and worker training. Locations with more conducive weather conditions would facilitate labor productivity more than those with regular and/or extended inclement weather conditions. Extreme weather conditions can slow down work progress or halt it altogether, thus resulting in labor productivity fluctuations.
8) *Labor relations, apprenticeship, and training*: The prevailing labor relations, apprenticeship, and training programs at a project location can affect labor productivity fluctuations. Well-trained and skilled workers are more efficient and effective in handling project tasks than their untrained or poorly trained counterparts. The available apprenticeship and training programs may vary due to the prevailing labor relations. For instance, some locations have union labor and nonunion or merit shop labor or both. Apprenticeship and training programs may not be equally available for the two labor pools. In theory, union labor may be more skilled, as they are organized and often tout themselves as offering robust apprenticeship and training programs, ensuring that their members have the right training and skills for the project work that they perform.

Labor productivity fluctuations, like those of equipment productivity, are among the most uncertain aspects of construction projects to estimate accurately and consistently. Both are influenced by a combination of internal and external factors, and proactive management strategies are crucial in mitigating associated challenges and maximizing productivity.

Labor productivity fluctuations can indeed be a cause of dispute in construction, and here are the reasons why:

1) *Project delivery timeline as contractual obligation*: The project owner and the contractor, by the terms of the contract, agree for the project to be completed within a stipulated time frame. If labor productivity fluctuation results in delays that extend the project completion date beyond what was agreed, accounting for justifiable adjustments, that would be a breach of the contract that can lead to disputes between the parties over who takes responsibility for the consequences and damages of delay.
2) *Cost overruns*: Labor productivity fluctuations resulting in delays can lead to cost overruns due to paying more for labor that is out of sync with expected productivity. Additional work hours, overtime, and extra hands may all be needed where productivity is slowed to accomplish the project. Cost overruns are a ready cause of disputes between the parties.
3) *Quality issues*: Labor productivity fluctuations can affect quality, where decreased productivity can lead to rushed work, cutting corners, and errors in handling increased workload while struggling to meet the deadline. Quality of workmanship and materials can be a major cause of disputes between the parties.
4) *Argument over liability and responsibility*: Where the causes of decreased labor productivity are external, such as inclement weather, regulatory enforcement, material delivery delays, design changes, and approvals, arguments may arise as to which party is partly or completely liable and responsible. The owner may argue that the contractor's project management capability is subpar at managing the fluctuations, while the contractor can claim that they are all beyond the contractor's control.

Overall, to mitigate and avoid disputes from labor productivity fluctuations, it is essential for the best practices of risk assessment and labor engagement to be embraced. The contractor must assess the project type, complexity, location, and the available labor pool, skills, and experience to accomplish the work. Based on gathering and analyzing these pieces of project information ahead of bidding, the contractor is able to proactively plan to handle labor productivity fluctuations to its advantage for a win-win for both parties.

Please refer to Table 14.1 for the recommended PDM's D-PAD concept applicable in this regard.

3.9 Equipment Productivity Fluctuations

Equipment productivity refers to its production rate, measured as the time taken to accomplish or produce a unit of work or product. For instance, with regard to excavating and related types of production equipment, the fifty-minute hour is applicable, and that means that, on average, the equipment is productive for a net time of fifty minutes per hour. Equipment productivity can fluctuate due to the following reasons:

1) *Equipment condition*: The condition of the equipment plays a crucial role in its productivity. Newer and well-maintained equipment is likely to perform efficiently and effectively, thus producing more with fewer breakdowns. Older and poorly maintained equipment will be the opposite.
2) *Operator skill and experience*: An operator who is skilled and experienced in using an equipment is likely to use it more effectively to produce more than one who has no such skill and experience.
3) *Operator attitude*: Some operators may delay more than others due to their attitude toward work. They may take unauthorized breaks, such as bathroom and smoking breaks.
4) *Nature of equipment operation*: Regardless of the condition of the equipment, some may require certain shifting or turning positions to accomplish a task. These shifting and turning positions fluctuate the productivity level.
5) *Downtime and maintenance*: Planned and unplanned maintenance, such as minor repairs, fueling, and greasing, all come with downtime for the equipment, and this can fluctuate and decrease productivity. Regular maintenance can help to minimize downtime from breakdowns and minor, on-site repairs.
6) *Industrial action*: Boycotts, picketing, and strikes by workers, as well as labor shortages, can affect equipment productivity due to no or less availability of operators.
7) *Material availability*: When the material or equipment is not available, the equipment will be idle, and productivity will be affected.
8) *Location and site condition*: Weather, terrain, and accessibility of the site can affect equipment productivity. Inclement weather conditions, such as heavy storms, snow, ice, and extreme heat, can slow down or disrupt equipment operation, and challenging terrain may make it difficult to maneuver some equipment, leading to decreased and fluctuating productivity.
9) *Work changes*: Work changes can affect equipment productivity, as some equipment may need to be reconfigured or repositioned to handle the changed work.
10) *Project management and scheduling*: Effective project management and scheduling are essential in coordinating and deploying equipment for optimal productivity. Poor scheduling and coordination can lead to inefficiencies and downtime, leading to decreased and fluctuating equipment productivity.
11) *Advancements in the equipment*: Adoption of new equipment and technologies demands some learning curves, which can affect productivity fluctuations until they are mastered.

By implementing strategies to manage and minimize the impact of these productivity factors, expected or estimated equipment productivity levels can be maintained.

Many dispute issues related to labor productivity fluctuations also apply to equipment productivity fluctuations. Equipment productivity fluctuations can be causes of dispute in construction due to the following reasons:

1) *Project delivery timeline as contractual obligation*: The project owner and the contractor, by the terms of the contract, agree for the project to be completed within a stipulated time frame. If equipment productivity fluctuation results in delays that extend the project completion date beyond what was agreed, and allowing for justifiable adjustments, that would be a breach of the contract that can lead to disputes between the parties over who takes responsibility for the consequential damages of delay.
2) *Cost overruns*: Equipment productivity fluctuations that result in delays can lead to cost overruns due to paying more for labor that is out of sync with expected productivity. Additional work hours, overtime, and extra hands may all be needed when productivity is slowed to accomplish the project. Cost overruns are a ready cause of disputes between the parties.
3) *Quality issues*: Equipment productivity fluctuations can affect quality when decreased productivity can lead to rushed work, cutting corners, and errors in handling increased workload while struggling to meet the deadline. Quality of workmanship and materials can be a major cause of disputes between the parties.
4) *Arguments over liability and responsibility*: Where the causes of decreased equipment productivity are external, such as inclement weather, regulatory enforcements, material delivery delays, design changes, and approvals, arguments may arise as to which party to attribute part or all of the resultant liability and responsibility. The owner may argue that the contractor's project management capability is subpar at managing the fluctuations, while the contractor can claim that they are all beyond the contractor's control.
5) *Work changes*: Equipment productivity fluctuation may necessitate work changes and even an expanded scope that would call for additional resources to maintain project schedules. Disputes can arise over the approval and payment for these changes caused by equipment productivity fluctuations, especially where there is a disagreement about the cause or responsibility for the fluctuations.
6) *Recordkeeping*: Equipment productivity is factored into the initial estimate and must be monitored and documented through the project control process as construction proceeds. Disputes over equipment productivity fluctuations often stem from contested documentation and evidence of fluctuations, delay

causes, and impacts. Poor recordkeeping and disagreements over the accuracy of productivity measurements and impacts can exacerbate conflicts and disputes between the parties.

Overall, to mitigate and avoid disputes from equipment productivity fluctuations, it is essential that the best practices of risk assessment, proactive equipment inventory, and the state of repair are embraced. The contractor must assess the project type, complexity, location, and the available equipment inventory, skills, and experience of the operators to accomplish the work effectively and efficiently. Based on gathering and analyzing these pieces of project information ahead of bidding, the contractor is able to proactively plan and handle equipment productivity fluctuations to its advantage for a win-win for both parties.

Please refer to Table 14.1 for the recommended PDM's D-PAD concept applicable in this regard.

3.10 Inclusion of Alternates

If allowed, contractors can include alternates in their bids for a project. Inclusion of alternates means that alternative materials and components are allowed within a project's specification or plan under certain conditions, such as the establishment of equivalency or equality. Alternates are typically substitute or additional materials and components proposed alongside the original design and specification or scope of work. Alternates are often included by the contractor due to the following reasons:

1) *Competitive bid*: Where qualified bids are allowed, the contractors may compete as offerors of a competitive bid by including alternative materials and innovative solutions that can offer quality as well as value for money.
2) *Value engineering*: Alternatives can be suggested in the value engineering process to optimize project value without sacrificing quality. In value engineering, the project's design features, structures, systems, equipment, materials, and components are analyzed to optimize the project outcome in terms of value for cost. Note that value engineering is not about cost-cutting, but maximizing value for the cost or the money invested.
3) *Cost management*: Inclusion of alternatives can be deployed for cost management of the project. Different cost scenarios can be explored, and the most cost-effective solution can be determined and used. Bids can be submitted on the different combinations of alternates, helping owners to understand the cost implications of certain design and material choices. Doing so helps to balance project quality goals with financial budgetary constraints.
4) *Flexibility*: Inclusion of alternates offers flexibility during the project construction process and allows for adjustments in cases of design and budget changes,

as well as material availability. For instance, Andersen windows may be what was specified, but as the construction got to window installation, the specified windows were not completely available; hence, Jeld-Wen windows were substituted as equivalent or equal to Andersen windows in terms of quality and performance expectations. As a result, the delay of waiting for Andersen windows to become available is avoided.

5) *Contingency*: Inclusion of alternatives can serve as a form of contingency for when certain components or elements of the project encounter unforeseen issues or cost overruns. Under such circumstances, alternatives may be resorted to as a backup option to keep the project within budget and schedule.

Essentially, inclusion of alternatives provides a means to optimize project outcomes by being flexible and considering different options and cost-value trade-offs throughout the construction process.

While the inclusion of alternates in the project construction process offers a lot of benefits, it can also bring about disputes between the parties due to the following reasons:

1) *Establishment of equivalency or equality*: When a design specification allows alternates by having product brand names include the phrase "or approved equal" or "or equal," it invites the need for quality checks to establish equality. This need is so even where full descriptions and technical data of the alternates are provided, together with a statement of the cost addition to, or deduction from, the base product bid, if approved. This scenario can occur in the above example, where Jeld-Wen windows were used to substitute Andersen windows. Disputes can arise between the parties as to the equality of Jeld-Wen and Andersen windows in cost, performance, and quality. While the contractor can argue that they are equal in every aspect, the owner may posit that cost should be lower, as Andersen is a noted brand and costlier than Jeld-Wen, even when their performance and quality may match equally.
2) *Contested bid award*: Where alternates are allowed for bidders, it means that qualified bids are welcome, and this may not provide a level playing field for all bidders. If there is no transparency or clear communication of how alternates are evaluated in the bid selection and award, it may lead to disputes.
3) *Quality issues*: While allowing contractors to include alternates may offer cost savings, it often invites the use of inferior-quality materials, and this can lead to disputes between the parties. Owners will dispute alternatives that compromise project quality, while contractors push for their inclusion to reduce costs.
4) *Cost escalation*: Where alternates come with discrepancies and costs additional to the base product bid if approved, they can accumulate to constitute cost overruns for the project, and this can bring about disputes between the parties.

5) *Pricing and timing of alternates*: If alternates are introduced due to unforeseen circumstances or changes in project requirements, their pricing and timing trigger disputes between the parties. Some owners may need time to adjust to them, and that can be a point of dispute if they extend the project timeline.

In order to prevent and/or avoid disputes from the inclusion of alternates, the parties should be proactively decisive on the design, materials, and components at the onset and not include them at all; if they must be included, parties need to be transparent and clear in communication, with alternate descriptions and data indicating whether their inclusions are cost additive or deductive to the base product bid.

Please refer to Table 14.1 for the recommended PDM's D-PAD concept applicable in this regard.

3.11 Underbidding and Accuracy of Estimates

Underbidding happens when a contractor bids for a project lower than the cost of completing it. Estimating involves projecting the costs of materials, labor, equipment, and other direct and indirect costs associated with the construction of a project. The accuracy of estimates relies heavily on a thorough understanding of the project scope, delivery process, market conditions, and potential risks. The accuracy of estimates is crucial to ensure that the project is delivered within the budget. If not deliberate or forced by economic circumstances, underbidding can be the result of inaccurate estimates, and both occur due to the following reasons:

1) *Desperation*: The contractor may be desperate to win the bid on a project during an economic downturn or slowed business, and it prioritizes securing the project over making a profit.
2) *Competition*: Competition can also motivate the contractor to underbid so as to beat competitors in a competitive market, and this may be done at the sacrifice of profitability.
3) *Mistaken calculation*: Inexperienced contractors and estimators can miscalculate estimates and complete an estimate full of omissions. They may not thoroughly evaluate the project requirements, overlook certain details, and/or fail to anticipate and account for potential complications.
4) *Optimistic judgment*: The contractor may be overly optimistic about its ability to cut costs and build the project cost-effectively, but the reality on site may make this optimism idealistic rather than realistic.
5) *Project scope definition*: If there is ambiguity in the project scope and it is not clearly defined, it may be difficult to accurately estimate and bid on it.

6) *Outdated historical data*: Contracts rely on historical project and estimate data in their ledger. If these data are outdated and are relied upon for current estimates, the accuracy cannot be guaranteed, and underbidding can happen as a result.
7) *Market condition fluctuations*: Market condition fluctuations for labor availability, equipment rates, and material pricing may be difficult to predict and accurately allow for in estimates, and this can lead to underbidding. Hence, the contractor must aspire to keep current with market trends to accommodate any changes in estimates and bids.
8) *Poor culture of risk assessment*: If the contractor is subpar in its risk assessment culture and strategies, it may be unable to proactively identify and quantify potential risks, such as unforeseen site conditions, possible severe weather delays, regulatory changes, and constraints. An estimate that fails to consider those may not be accurate and can lead to underbidding.
9) *Expertise and experience*: Contractors with the necessary expertise and experience in estimating and bidding will likely not inaccurately estimate and underbid when compared with those lacking such exposure.

Underbidding from inaccurate estimates can be very consequential for both the contractor and the owner. For the contractor, it will bring financial losses and possibly lead to corner-cutting and lowering of quality to reduce costs and remain within budget. Such actions would compromise the quality and integrity of the project.

Underbidding and inaccuracies in estimates are common causes of disputes between the parties due to the following reasons:

1) *Owner expectation*: Underbidding and inaccuracies in estimates can raise unrealistic expectations in the owner about the cost and timeline, which, if exceeded in the end, would lead to disputes over the feeling of being misled by the contractor.
2) *Compromised quality*: When a contractor inaccurately estimates and underbids a project, it may resort to corner-cutting and the use of inferior materials and components in order to keep costs down and remain within budget, but this can bring about disputes with the owner.
3) *Manufacturing of change orders*: A contractor who has inaccurately estimated and underbid a project may resort to unethical practices of generating unwarranted change orders as a way to cover the cost of additional and unforeseen work, but the owner may argue that the contractor should have foreseen and accommodated them in their estimates and bid. This disagreement is certain to result in disputes.
4) *Cost overrun*: If the actual cost exceeds the estimated and bid amount, an argument can arise as to which party should bear the burden of the additional cost.

The owner may feel misled by a desperate contractor who might have cooked estimates and submitted a low bid just to win the job with an agenda to increase and demand compensation as the project progresses.

5) *Scope creep*: When the estimates and bids are based on a poor or a lower comprehension of the project scope, the reality on site may present a bigger scope, leading the parties to dispute over the extra work, time, and resources, and who should bear the costs.
6) *Schedule disruptions and project delays*: The project timeline runs over if the contractor misinterprets the scope and underestimates the time, materials, equipment, and resources required to complete the project within budget and schedule. Consequential damages are a major cause of disputes between the parties.

In order to prevent and/or avoid disputes from underbidding and inaccurate estimates, the contractor must exert maximum expertise and experience in thoroughly analyzing the project and assessing it for all potential risks that may lead to inaccurate estimates and underbidding. Additionally, using unambiguous contracts, clauses, and strategies to address any change orders and unforeseen circumstances, and clearly conferring and communicating effectively with the owner, can avert disputes during the construction process.

Please refer to Table 14.1 for the recommended PDM's D-PAD concept applicable in this regard.

3.12 Unrealistic Scheduling

Unrealistic scheduling of projects occurs when the contractor sets timelines of delivery that are overly optimistic or ambitious, which are not feasible considering the project realities of scope, available resources, and potential challenging and unforeseen circumstances that may arise as the project progresses. Depending on experience, three types of timeline estimates are possible on projects: (1) optimistic estimate which reflects or assumes ideal circumstances; (2) most probable estimates which accounts for and accommodates usual contingencies such as routine delays; and (3) pessimistic estimate which reflects a worst-case scenario based on Murphy's Law and the assumption that everything that can go wrong will go wrong. Here are some of the reasons why unrealistic scheduling can occur in construction:

1) *Pressure to win bids*: In competitive markets, the contractor may submit an aggressive but unrealistic and unfeasible project schedule just to win the bid over competitors. Then the contractor muddles through and struggles to meet the project timeline.

2) *Lack of expertise and experience:* The contractor may be working with unrealistic scheduling, mainly due to a lack of expertise and experience. Inexperienced contractors, estimators, and schedulers may underestimate the time and resources required to complete aspects of the project. Without expertise and a thorough understanding of the intricacies of the project's construction process, they may set unrealistic and unfeasible deadlines and milestones.
3) *Oversight of potential delays:* Delays occasioned by adverse weather conditions, materials supply chains, regulatory enforcements, inspections and approval constraints, and unforeseen site conditions and project aspects are commonplace in construction projects. Unrealistic scheduling by project planners and schedulers can occur when these routine delays are not anticipated and accounted for when setting project timelines.
4) *Overly optimistic assumptions:* Project planners and schedulers can make overly optimistic assumptions about potential adverse weather conditions, materials supply chains, regulatory enforcements, inspections and approval constraints, and unforeseen site conditions and project aspects. These assumptions can result in unrealistic scheduling when these factors turn out to be other than the ideal anticipated.
5) *Pressure to meet deadlines:* The contractor may be under pressure from the owner, investors, or stakeholders, and therefore resort to aggressive but unrealistic and unfeasible schedules to satisfy them at the risk of not fully considering the complexities and the nature of the project at hand.

Unrealistic scheduling can be vastly consequential, and it can damage the reputation of the contractor and lead to disputes with the owner in several ways:

1) *Missed deadlines*: When working with unrealistic scheduling, it means that the deadlines and milestones are spurious and prone to being missed. Disputes will arise with the owner, investors, or stakeholders who are working with agreed deadlines, which are unrealistic and unfeasible after all.
2) *Compromised quality*: The contractor working with unrealistic scheduling may be compelled to cut corners and rush some aspects and phases of the project in order to meet the project timeline, and this can compromise workmanship and project quality, which would lead to a dispute with the owner.
3) *Scope creep and change orders*: As the contractor might have overlooked critical aspects of the project in setting an unrealistic schedule, the reality of the project at the site can mean increased scope and change of work, and disputes can arise over which party would bear the responsibility of the additional costs.
4) *Resource constraints*: Unrealistic scheduling can stretch and strain available resources of labor, equipment, and material. Besides the owner, this situation can bring about disputes among project participants, such as subcontractors,

suppliers, and vendors over inadequate resourcing to meet the requirements of the project.

5) *Breach of contractual obligations*: If a contractor commits to delivering a project within a certain contractually agreed time frame, albeit unrealistic and unfeasible, but fails to meet that deadline, the owner can claim consequential damages and penalties as specified in the contract. The contractor can dispute that unrealistic expectations were imposed on them or that inaccurate information was provided during the bid process, leading to unfeasible scheduling, resulting in time extensions beyond what was agreed in the contract.

In order to prevent and/or avoid disputes from unrealistic scheduling, the contractor must exert maximum expertise and experience in thoroughly analyzing the project complexity and intricacies, analyzing it for all potential risks of delays from known and unknown factors that often occur during construction. This will help the contractor to set realistic schedules based on a thorough understanding and planning, accurate assessments of project circumstances and requirements, and consideration of potential project challenges and contingencies. Effective communication and a collaborative approach can also help to avert disputes between the parties.

Please refer to Table 14.1 for the recommended PDM's D-PAD concept applicable in this regard.

3.13 Project Resourcing and Balancing—Labor, Materials, and Equipment

Project resourcing and balancing refer to the management and coordination of human, material, equipment, time, and financial resources to ensure efficient and effective project delivery within budget, on schedule, and to quality. The process involves:

1) *Human resource management and coordination*: Projects call for different expertise, skills, and trades, and a contractor that knows his stuff can figure out the right mix of skills and trades to accomplish the various project tasks efficiently and effectively. The contractor should ensure neither to over-resource a particular task nor to under-resource others—that is, balancing from a labor perspective.
2) *Material resource management and coordination*: Construction projects involve a lot of materials, from raw to finished products. Efficient and effective material resource management is needed to ensure materials' availability when needed without delays or extra costs. Material resource management and coordination

involve coordinating with suppliers and vendors, managing material inventory, and optimizing its usage.

3) *Equipment management and coordination*: Efficient and effective equipment management is needed to ensure its availability when needed without delays, unplanned downtimes, or at more than estimated costs. Equipment management involves ensuring that it is in good repair and is handled by people with expertise and skills to operate effectively and efficiently.
4) *Financial management*: Effective financial management is all about keeping the project within the budget by appropriately allocating and using finances where and when needed. The contractor should ensure neither to over-finance a particular task nor to under-finance others—that is, balancing from a financial perspective.
5) *Time management*: Construction projects often work with tight deadlines. Hence, time is finite and project tasks must be scheduled so as to ensure that they are accomplished on time and not delay other parts, jeopardizing meeting the project timeline. Time management requires strategic allocation of resources, isolating and prioritizing critical tasks, and applying tested and trusted project time management techniques, such as the critical path method (CPM) or program evaluation and review technique (PERT).
6) *Risk management*: Construction projects are inherently risky, with potential challenges from adverse weather, unforeseen site conditions, and project circumstances. Effective resource management and coordination involve identifying a project's potential risks and figuring out a way to prevent, avoid, or navigate around them in the project delivery process. This may involve having contingency plans in place, such as alternative suppliers and backup resources.
7) *Collaborative communication*: Effective project resourcing and balancing cannot happen in the absence of clear and collaborative communication among participants and stakeholders. This ensures that all would work together to accomplish the shared or common project goals.

Essentially, project resourcing and balancing in construction require strategic planning, management, and coordination to efficiently and effectively achieve the project objectives within budget, on schedule, and to quality.

Due to the complexity and dynamic nature of resource management, it can indeed be a source of dispute in construction, and here are the ways that can be so:

1) *Equitable and efficient allocation of resources*: Where human/labor, material, time, and financial resources are perceived not to have been equitably and efficiently allocated by the contractor, friction may develop among project participants, leading to disputes.

2) *Disruptions and delays*: Poor resource planning and coordination can lead to project disruptions and delays, which, if resultant schedule changes result in timeline extensions, can become a dispute issue between the parties over who takes responsibility for the consequential damages.
3) *Cost overruns*: Inefficient allocation and use of resources resulting from poor resource planning and coordination can lead to unexpected expenses related to resource shortages, delays, or changes in project scope. Disputes may arise between the parties as to who should bear the responsibility for the extra costs and whether the extra expenses are justifiable.
4) *Compromised quality*: Poor resource planning and coordination can lead to avoidable cutting of corners and misuse of resources to cover up missteps or oversights. Substandard materials and engagement of less-skilled or unskilled labor can occur in this scenario, and it can lead to subpar, defective work and rework. Disputes can arise between the parties as to the responsibility for the cost of correcting defective work.
5) *Subcontractor and supplier/vendor disputes*: Project resourcing and balancing involve subcontractors and suppliers. Disputes may arise over material and equipment availability and delays, as well as payment.
6) *Lack of collaborative communication*: Ineffective communication and collaboration among project participants breeds misunderstanding and friction about resource needs, availability, and scheduling. This can escalate into disputes.
7) *Regulatory compliance and safety concerns:* Poor resource planning and coordination can come in the form of compromised standards and safety that can attract fines, disruptions, delays, and work stoppages. Understaffing and insufficient PPE can strain the workers and compromise safety measures that pose safety and health risks to the workers, potentially triggering litigable disputes.

Preventing, avoiding, or dousing these sources of disputes would require proactive strategic planning, clear and effective communication among project participants, and a collaborative problem-solving approach. Clearly defined roles and responsibilities, regular and strict project control and resource monitoring, and swift dousing of flaring frictional issues all contribute to proactively handling potential disputes.

Please refer to Table 14.1 for the recommended PDM's D-PAD concept applicable in this regard.

3.14 Use of Union and Nonunion Labor

Union labor or trade union labor refers to the collective organization of workers of a given trade or industry with the intent to better their welfare—collective bargaining for better wages, benefits, safety, and job security. On the other hand, nonunion labor refers to where workers are not organized into one bargaining or collective entity. The construction industry utilizes both labor forces, and you often hear "union contractors," meaning that the contractors use union labor and hire from union halls. Nonunion contractors hire from open or merit shops for nonunion workers. A contractor can use either or both workforces, and each has implications for cost-effectiveness and quality.

The features of union labor are as follows:

1) *Organization*: Workers are organized and represented as an entity in bargaining and negotiating contracts for their members in terms of work conditions, wages, benefits, and other employment entitlements and treatments.
2) *Standardized wages and benefits*: Union workers enjoy standardized wages and benefits as union contracts level on the number and capacity of membership to bargain and negotiate better standardized wage rates, work hours, and employment conditions and welfare (such as health insurance, retirement plans, and paid leave) across projects.
3) *Collective bargaining power*: Unionism gives workers a common, powerful voice based on their number in negotiations with employers, and unions are able to extract better wages and employment conditions in a collective, organized, and standard manner in what could otherwise be chaotic if left to individual workers.
4) *Organized training*: Unions often organize training and apprenticeships for their members, guaranteeing that they upskill to do their jobs satisfactorily.

The features of nonunion labor are as follows:

1) *Flexibility for employers*: When workers are not unionized, employers negotiate with each worker independently and determine their wage rates, benefits, and conditions of employment based on their merit or skills and experience—that is why they are called open shop or merit shop workers.
2) *Less stringent work rules*: Other than those required by law, nonunion workers work according to the work rules of the employer, thereby allowing for more adaptability and efficiency.
3) *Wage variations*: As individual workers negotiate for themselves, their wages may vary according to their bargaining power, skills, experience, and demand.
4) *Lower labor cost*: Construction projects using nonunion workers may have lower labor costs, as the employer decides their pay and benefits, and has leverage to make this most favorable for themselves. Such leverage can be helpful on projects with tight budgets or those that are competitive with competitors.

In practice, whether to use union or nonunion labor depends on various factors, such as project type and size, location, government regulations, budget, labor market conditions, and the preferences of the contractor or project owner. Some construction projects, especially government projects, may mandate the use of union labor, and if union labor is not used, the wages and conditions must be at par with the prevailing union wages and conditions in line with the Davis-Bacon Act of 1931. Others may opt for nonunion labor to maintain cost competitiveness and flexible labor hiring and management.

Union and nonunion labor play a crucial role in the construction industry, and each comes with different pros and cons for the stakeholders. Some of the cons, which can be sources of disputes between the parties:

1) *Clash of union and nonunion workers*: Wages and benefits disparities between union and nonunion workers can be a source of disputes, as the former tend to fare better than the latter. Tensions may arise where both workers are engaged on a project, as nonunion workers may feel unfairly used and compensated for the same type of work.
2) *Budget management problem*: Where a construction project is mandated to use union labor, the contractor will not have the same flexibility to manage labor costs as it would with nonunion workers. A budget management problem that leads to extra costs can bring about disputes between the parties over the responsibility for additional costs.
3) *Project delays and productivity*: Union labor can embark on industrial actions such as picketing, strikes, and boycotts as they seek to negotiate better work conditions, wages, and benefits, and this can delay project progress. It also takes time to resolve unfair labor practices if and when the National Labor Relations Board's oversight is involved. Consequential damages and responsibilities are certain triggers of disputes between the parties.
4) *Quality issues*: The use of nonunion workers with their various skills and experience, and because the employer can pay them relative to their skills and experience, may motivate the employer to use the least skilled personnel. This can reflect in the workmanship and quality of their work. Poor workmanship and quality are sure triggers of dispute between the parties.

Overall, the exclusive or mixing of union and nonunion workers on a construction site comes with challenges that can trigger disputes. In order to avoid these disputes, contractors and owners should be clear on the type of workers (union or nonunion) to engage on the project, anticipate and proactively address any issues, and communicate effectively throughout the process.

Please refer to Table 14.1 for the recommended PDM's D-PAD concept applicable in this regard.

3.15 Work Quality–Omissions, Defects, Failures, Rework, and Punch List

The triumvirate of construction project management success is predicated on cost, quality, and time. Work quality in construction refers to the degree to which the project outcome fulfills the design intent as per the contract documents—drawings, specifications, contract, addenda, and change orders. Delivery work quality on a construction project is an expressed and implied fulfillment of the contract. In order to deliver work quality, the following are involved:

1) *Materials used and conformance with specifications, codes, and standards*: All materials used for the project must meet the job specifications, codes, and standard requirements.
2) *Workmanship*: The labor engaged must be of a skill and craftsmanship as to properly install and assemble the materials and components to deliver a quality project to professional standards.
3) *Errors and omissions*: Work quality that is high is likely not riddled with errors and omissions or rework. When the right skilled hands are used, errors and omissions as well as rework are reduced or eliminated. Attention to detail is likely in this scenario.
4) *Integrity of construction work*: Work quality that is high goes with durability and longevity because appropriate materials are used, and they are properly assembled or installed.
5) *Accidents and safety*: Work quality that is high is likely to be guaranteed with the right skilled labor and workmanship with the know-how of safe materials and components assembly and installation.
6) *Quality assurance, control, and management*: Work quality that is high is likely to be a result of quality systems in place or use that include quality assurance, quality control, and total quality management—QA/QC/TQM. Implementing QA/QC/TQM ensures that issues are identified, isolated, and addressed early or swiftly in the project process, ensuring a quality outcome that meets or exceeds expectations.
7) *Fitness for purpose*: Work quality that is high speaks to fitness for purpose. By implied warranty, a quality project that is well-constructed should be fit for its intended uses.

In summary, attaining a high work quality in a construction project involves engagement of the right materials, skills, and workmanship to plan and execute it to standards with adherence to excellence and professionalism.

Where work quality is in question, disputes can arise between the parties due to the following reasons:

1) *Breach of contract*: Work quality other than that adhering to the contract documents (drawings, specifications, contract, addenda, and change orders) constitutes a breach of contract, which is a known cause of disputes between the parties. Due to differing quality interpretations, disputes can arise over whether the delivered work quality meets the contractual specifications and standards.
2) *Negligent and defective work*: Work quality that is riddled with negligence, errors, and omissions, as well as rework, is certain to engender disputes between the parties. Improper assembly and installation of materials and components is negligent and defective.
3) *Cost overruns*: Reworking and correcting negligent and erroneous work and omissions costs extra, and cost overruns in a construction project are ready causes of disputes between the parties.
4) *Delays and damages*: Reworking and correcting negligent and erroneous work and omissions takes extra time, and exceeding delivery deadlines in construction projects is a ready cause of disputes between the parties due to the consequential damages and who is to take responsibility for the costs and liabilities.

Essentially, ensuring quality work through proactive and active steps akin to QA/QC/TQM would help to prevent, avoid, or douse quality-related disputes between the parties. This ensures the planning, execution, and delivery of quality projects that meet contractual specifications, codes, and standards.

Please refer to Table 14.1 for the recommended PDM's D-PAD concept applicable in this regard.

3.16 Workers' Fidelity

Workers' fidelity refers to the integrity and honesty of the employees engaged in the construction project. Are they time thieves (stealing or claiming unworked minutes and hours here and there), drug abusers, skimmers, larcenists, embezzlers, intellectual thieves, data thieves, harassers (sexual and unsexual), defamers and slanderers, etc.? Hence, workers' fidelity can show an impact on the projects in the following ways:

1) *Punctuality and presence*: Reliable workers always show up and are on time. You can count on their productivity relative to what was estimated at the project bid stage. They recognize the importance of their presence and punctuality to maintain the project's productivity in terms of schedule and costs. They don't pilfer off minutes and hours here and there and accept payment for their time and work, nonetheless.

2) *Drug-free and safe worksite*: Workers with integrity recognize the need to rid the project site of drugs and banned substances—weed, alcohol, etc. Such riddance translates to better worker and worksite safety from accidents and incidents and adherence to safety practices and protocols.
3) *Accountability*: Larceny is a problem in construction. Honest workers don't steal materials from the worksite. According to the U.S. Chamber of Commerce, three out of four employees admit to stealing from their employers at least once; one of every three business failures is the direct result of employee theft; and employee dishonesty losses incurred by American businesses total more than $50B annually (The Surety & Fidelity Association of North America: https://surety.org/surety-fidelity/what-is-fidelity/#learn-more, accessed 24 May 2024). Construction is one of such businesses.
4) *Prompt payment to workers*: Dishonest trades and subcontractors can embezzle or unreasonably delay payment to their workers and suppliers even after receiving the same from general contractors (GC). General contractors can also do the same to subcontractors and suppliers. These issues can affect the morale and productivity of the workers beyond conflicts and disputes. It is not surprising that some contracts come with pay-when-paid clauses in recognition of this reality.
5) *Harassment*: Harassment of all types can take place on the jobsite, and resources of time and money may be expended to deal with it. Sexual harassment, intimidation, and molestation can all happen on the job with workers of questionable integrity and decency.

Overall, workers' fidelity is crucial to the smooth and successful execution of the project, as it ensures a conducive work atmosphere, productivity that you can rely on, and trouble-free delivery.

When workers engaged on the project have questionable fidelity, they can become triggers of disputes between the parties in the following ways:

1) *Material and component deficits and losses*: When materials and components for the project are pilfered by the workers, it affects the work in terms of delays, extra costs to replace them, and a tense and suspicious work environment. This scenario can trigger fights, conflicts, and disputes between the parties.
2) *Compromised safety*: If workers are involved in the abuse of substances, and their adherence to safety practices and protocols is subpar, Occupational Safety and Health Administration (OSHA) enforcement can delay and cost the project, and such extra time and costs are potential triggers of disputes between the parties.
3) *Prompt payment*: When a pay-when-paid clause is in the contract, but not strictly adhered to by the GC and/or subcontractors in withholding or delaying

payment, this can bring about a lien on the project and disputes between the parties.

4) *Payment for time and work not done but claimed by workers*: When workers are time thieves by clocking minutes and hours not worked, the estimated labor productivity and cost will not add up. Hence, corners have to be cut to still meet the project schedule and budget, and it can expose the contractor to disputes with the owner.
5) *Harassment costs*: When workers indulge in harassment that distracts from the project objectives, resources are wasted in dealing with the consequences, and besides damaged reputation, corners may be cut to make up for the time lost and costs incurred, and that can lead to disputes between the parties.

In order to prevent or avoid disputes from workers' fidelity, contractors should ensure a proactive and thorough vetting of workers engaged on their job sites, verify past employment, provide clear policies on fidelity, set up a system of checks and balances, conduct random audits, maintain workers' morale, and take out a fidelity bond.

Please refer to Table 14.1 for the recommended PDM's D-PAD concept applicable in this regard.

3.17 Accidents and Safety—Experience Modification Rating

Accidents and incidents occur on a jobsite where safety practices and protocols are not obeyed or are nonexistent. Safety is a moral and ethical issue with immediate, direct, and indirect costs from accidents and incidents affecting the reputation, competitiveness, and survival of the contractors. The direct costs from damage to the project and workers' compensation are often covered by insurance, but the direct costs pale in comparison to the indirect costs such as downtime, halted work, scope and schedule creep with consequential damages, new hiring and training, as well as negative publicity and loss of competitiveness from higher insurance premiums due to higher experience moderation rating (EMR). EMR is used by insurance companies to calculate accident and incident frequencies for contractors on their jobsites, and they use this rating to moderate the premiums they pay for workers' compensation insurance.

Accidents or incidents can come from falls (such as falling from heights, scaffolds, ladders, or roofs), explosions, exposure to toxic chemicals, electrocution, structural collapse, being struck by objects, being caught in between objects, fire outbreaks, drugs, and alcohol. There are many types of recordable accidents and injuries on a project site, as classified by OSHA, and they include:

1) Death—someone dying on the jobsite.
2) Illness leading to days away from work.
3) Incident leading to a restriction on usual work, or a transfer to another work due to the limitations from the incident.
4) Incident that results in medical treatment beyond first aid.
5) Incident leading to loss of consciousness.

These accidents and incidents are caused by:

1) *Human unsafe acts*: This is the topmost cause of accidents, and they come from human attitude and disposition toward safety, errors, negligence, fatigue, distractions, and lack of training.
2) *Unsafe work conditions*: Lack of established safety best practices and protocols, such as safety signage (e.g., Job Safety and Health: It's the Law!), hazard communication protocols like safety data sheets (SDS), poorly maintained equipment, and use of barriers—all have an impact on the occurrence of accidents and incidents on the jobsite.
3) *Weather conditions*: Weather conditions such as rain, storms, snow, and ice can all create hazardous conditions for a jobsite.

Overall, dealing with accidents and incidents on a jobsite is essential to the project outcome. Where safety measures—training of workers on construction techniques, equipment used, and the safety protocols; providing and enforcing the use of PPE such as hard hats, harnesses, gloves, proper clothing, and safety glasses; regular inspections; proactive safety risk assessments; hazard identification and communication through "haz com" such as SDS; emergency preparedness; and strict compliance with OSHA regulations—are not taken to prevent them, they can lead to disputes between the parties, and here's how:

1) *Adherence to contractual obligations*: Disputes can arise between the parties over liability and contractual compliance if accidents occur, as contracts typically outline safety requirements and responsibilities of the parties.
2) *Scope and schedule creep*: Accidents and incidents on a project can halt or delay it to the extent that more resources of time, labor, and equipment will be required for timely delivery. Such extra resources are triggers for disputes between the parties.
3) *Workmanship*: Work quality can be affected by accidents that lead to the hiring of new workers if accidents and incidents put the original workers out of work. When the project quality is affected, disputes can arise between the parties.

4) *Tense work environment and strained relationship*: A worksite riddled with accidents and incidents brews a tense and contentious environment that can easily trigger disputes between the parties, besides the potential to strain the relationships.

In order to prevent, avoid, or defuse disputes from accidents and safety issues, the parties must prioritize safety and adhere to OSHA regulations and safety best practices on the jobsite. Any safety concerns must be swiftly addressed, and any potential disruptions to the project's progress must be forestalled immediately. A safety culture should be promoted where all workers feel empowered to report hazards and suggest or proffer immediate solutions as necessary. Management must show an outward commitment to safety, buttressed by robust safety policies and proactive communications, regular safety meetings like toolbox talks, and emphasizing the critical importance of safety on construction sites.

Please refer to Table 14.1 for the recommended PDM's D-PAD concept applicable in this regard.

3.18 Mechanics Lien

Mechanics' lien refers to a claim a contractor or subcontractor has over an owner's project or property as a result of nonpayment for work done or materials supplied. Mechanics' lien allows a foreclosure action, forcing impounding or sale of the property in lieu of compensation. As discussed earlier under the topic "Distinguishing PDM from Partnering, Alliancing, IPD, ADR & Litigation," the doctrine of privity of contract in common law provides that a contract cannot confer rights or impose contractual obligations on someone who is not a part of or a party to that contract. However, the mechanics lien law extends lien rights up to the second-tier relationships and limits them beyond this point. It means that the contractor and the subcontractor can lien on the owner's project or property, but not sub-subcontractors or others. Sub-subcontractors can lien on the property of the contractor and the subcontractor, but not on the owner's property. Mechanics' lien works in the following ways:

1) *Work performed or materials supplied*: The contractor or subcontractor performs the work, and the supplier supplies the materials.
2) *Payment not made or received within agreed time frame*: If payment for work performed or materials supplied is not received within an agreed time frame, the claimant has a right to file a lien on the property, which is the subject of the work performed or the materials supplied.
3) *Notice of intent to file a lien:* Many jurisdictions require a notice to be given of an intent to file a lien due to nonpayment for work done or materials supplied within a specified period of time.

4) *File a lien*: In the absence of payment within a specified time frame, the lien will be filed. The advice of an attorney can be helpful while filing a lien. The timing to file a lien differs from state to state in the United States, but contractors typically have 90 days after the physical completion of the work or the project that is the subject of the lien. Note that a lien cannot be filed against public or state property, whether under the New York or Pennsylvania lien statute system.[1]
5) *Lien enforcement*: After a lien is filed, and no payment is received, proceedings must be brought to enforce the lien, usually within 90 days after filing. The services of an attorney will be required, as enforcement involves a foreclosure action. Lien laws are based on the statute of fraud or the theory of unjust enrichment against the unpaid party.

Mechanics' lien serves as a powerful instrument to protect contractors, subcontractors, and suppliers, and it ensures that the project owner pays promptly and fairly for work done and materials supplied when due. Mechanics' lien can indeed be a cause of disputes in construction projects, and here are the reasons why:

1) *Nonpayment for work done and materials supplied*: Disputes may arise over nonpayment, underpayment, unfair payment, or delayed payment due to work done and materials supplied that are not up to the agreed-upon quality or standards. The validity of the lien claim can be challenged.
2) *Lien process*: Filing a lien follows a process starting with a notice of intent to file a lien, and filing the lien itself, each with its timing. When the specified timing is not followed, disputes may arise as to the loss of the right to file a lien.
3) *Discrepancies and disagreements*: Disputes can arise over documentation and records of work completed, materials supplied, and payment made or received. Discrepancies can lead to a challenge to the validity of the lien.
4) *Legal proceedings*: Where the unpaid party gets no payment 90 days after filing a lien, the next step is legal proceedings to enforce the lien, which in itself constitutes a full-blown dispute that would add extra time and costs to the project.

Essentially, a mechanics lien has the potential to trigger and escalate disputes in a construction project. In order to prevent, avoid, or defuse possible disputes, lien waivers should be proactively explored, and if not, there should be careful documentation, clear and consistent communications, buttressed by prompt and

1 Under the New York lien statute, the amount of payment to collect by a lienor (subcontractor or supplier, for instance) is limited to the sum due to contractor from owner, and nothing will be collected by lienor from owner if no more sum is due to contractor from owner. However, under the Pennsylvania lien statute, a lienor subcontractor or supplier can collect sum from owner even if the entire contract sum has been paid to contractor.

fair payments for work done and materials supplied. Any payment issues should be swiftly resolved to defuse the possibility of disputes.

Please refer to Table 14.1 for the recommended PDM's D-PAD concept applicable in this regard.

3.19 Bonds

Bonds, appropriately called surety bonds, are important in construction as a guarantee of a successful project outcome. From bid bonds to contract bonds (namely, performance and payment bonds), many types of bonds exist to facilitate a successful construction delivery process. Bonds are provided by surety companies that rate contractors and levy premiums according to capacity, capital, and capability. Contractors are given credit ratings much like those for individuals with Equifax, TransUnion, and Experian, or like companies and countries with Standard and Poor's (S&P), Moody's, and Fitch Ratings. Contractors with an A1 rating are considered high-quality with the lowest credit risk; those rated A are considered upper-medium grade or quality with low credit risk; those rated B are considered medium grade with moderate credit risk; those falling into the miscellaneous category are a combination of two or more ratings in one. Here are some of the bonds most in use in the building and construction industry:

1) *Bid bond*: This bond is taken out by contractors to guarantee the project owner that they stand behind their bid sums and will enter into a contract if awarded, and provide performance and payment bonds as needed.
2) *Contract bonds*: Contract bonds refer to performance and payment bonds.
   a) *Performance bond*: This is the type of bond taken out by contractors, guaranteeing to the project owner that they will perform all required tasks in the successful delivery of the project according to the contract. This bond protects the project owner, in that the surety ensures that the project tasks are performed by the contractor, and if not, the surety company will take steps to do so and cover all and any costs.
   b) *Payment bond*: This is the type of bond taken out by contractors, guaranteeing to the project owner that they will pay for all labor, equipment, and materials required in the successful delivery of the project according to the contract. This bond protects third parties who are not in the privity of the contract.
3) *Subcontract bonds:* This is a replication of bid, performance, and payment bonds between the contractor and subcontractors, but the contractor is the obligee here. This bond protects the contractor, and its bid bond operates under the doctrine of equitable estoppel, guaranteeing that the subcontractor cannot estop on the contractor after the latter has relied on the former's numbers to submit and win its bid.

4) *Supply bond:* This is the type of bond required from materials and components suppliers and vendors, guaranteeing that the materials and components for the project will be timely delivered in accordance with agreed-upon terms.
5) *Maintenance or warranty bond:* This bond guarantees to the project owner that the contractor will fix at its own cost all defects in material and workmanship that become apparent after the completion of the project beyond the warranty period, which is often one year.
6) *Miscellaneous bonds:* Many other types of bonds may be required and provided in construction:
   a) *Environmental bond:* This bond guarantees that the contractor will comply with all environmental regulations and see to any required abatement or remediation.
   b) *Bond to release retainage:* This bond can be used to exchange for the amount of money held in retention.
   c) *Bond to discharge lien or claim:* This bond is provided in exchange for the discharge of a lien or a claim.
   d) *Bond to indemnify owner against lien:* This bond may be required from the contractor by the owner to protect against any lien on the latter's project or property.
   e) *Bond to protect owners of rented equipment or leased property:* This bond ensures that equipment renters pay rent, can cover damages beyond normal wear and tear, properly return equipment that is leased or rented, and offers financial recourse when the renter defaults.
   f) *Fidelity bond*: This bond protects the contractor against employee theft or dishonesty.
   g) *Judicial or court bond*: This bond guarantees or covers lawsuits that may arise during the project delivery process.
   h) *License or permit bond*: This bond covers any license or permit violations.
   i) *Termite bond*: This bond is given by applicators or manufacturers of termite applications.
   j) *Subdivision bond*: This bond guarantees the construction of all service utilities.
   k) *Self-insurer worker compensation bond*: This bond guarantees the payment of the employees of a self-insured contractor.
   l) *Union wage bond*: This bond guarantees the payment of union wages and benefits on the project.
   m) *Bonding alternatives*: Personal or individual sureties in the form of U.S. Treasury bonds or notes, certified or cashier's checks, bank drafts, postal service money orders, irrevocable letters of credit, or cash can be used in lieu of bonds if agreed.

Essentially, bonds play a crucial role in ensuring that projects are delivered successfully with all engaged labor, equipment, and materials paid for. However, bonds can become a source of disputes in the process in the following ways:

1) *Inordinate bond requirement by owner*: The project owner may make onerous bond requirements from the contractor, ranging from the scope, cost, and selection of bonding companies. They may request too many bonds that can stretch the contractor, or demand that the bond be taken out from designated surety companies of the owner's choice. These demands may lead to disagreements and disputes over the scope, type, and cost of the bonds.
2) *Bond costs*: Where the premiums of all the requested bonds are perceived to be excessive, arguments and disputes may arise as to which party would cover all or a part of the costs.
3) *Frivolous bond claims*: A party may unilaterally believe that the contract has been breached by the other party and can proceed to claim against the provided bond. The validity of such a claim can be disputed by the affected party.
4) *Disparity in performance perceptions*: One party may believe that performance is subpar and proceed to claim against the provided performance bond, while the other party can dispute the other's performance assessments and the severity as to warrant a claim.
5) *Payment disputes*: When subcontractors file a claim against a payment bond, it means that the contractor has not made timely payments to them as per the pay-when-paid contract clause. This occurrence can be a trigger of dispute between the parties.
6) *Bond coverage*: The worth of the bond is no greater than the surety's ability to pay. Hence, the owner often approves the surety company to use and the form of bonds to provide. Nonetheless, in cases of defects, delays, and other issues covered or assumed to be covered by the bonds, disputes may arise as to the differing interpretations of the bond coverage.
7) *Bond release dispute*: Upon project completion, disputes may arise as to whether the conditions of the bond have been met as to warrant its release: for instance, satisfactory completion of the works, fixing of all defects, and resolution of all outstanding claims.
8) *Bond conditions enforcement*: A dispute may arise over the modalities to fulfill bond conditions. For instance, in a performance bond, the surety company can get the contractor to perform, or engage others to perform where the contractor cannot, or pay the owner to engage others to perform where the contractor cannot. The procedures of filing and resolving the claims can be very contentious and disputatious.

In order to prevent, avoid, or defuse disputes from bonds, it is essential for the parties to clearly define the required bonds, the coverages, and the procedures of claim and resolutions if and when applicable. Where bonding issues are not preventable or avoidable, open communication must be maintained between the parties to swiftly defuse any possibilities of disputes that will take time and cost extra on the project.

Please refer to Table 14.1 for the recommended PDM's D-PAD concept applicable in this regard.

3.20 Insurance

Construction is a risky business, and players can quickly go out of business. Hence, risk management is essential in the industry. Risk management is a comprehensive approach to handling exposures to losses and perils with financial costs to the business. One of the key risk management approaches is insurance. Insurance refers to a contractual agreement between the insurer and the insured whereby the former assumes the financial risks and responsibilities of the latter for specified losses or perils upon the payment of associated premiums. Financial losses and perils can quickly bankrupt a business, but insurance companies step in to assume the risks, ensuring business continuity despite the loss of peril. Here are some common types of insurance in the construction industry, some of which are mandated by law:

1) *Workers' compensation insurance*: This is a mandatory insurance that covers medical expenses and lost wages for an injured or sick employee on the jobsite. It protects the contractor from lawsuits associated with injuries on the jobsite. With regard to possible injuries to workers on the project, they may be temporary total (injury or illness that knocks down the worker out of the jobsite for the short-term, and the employee can return to usual work after; e.g., a headache or dizziness), permanent partial (injury or illness that knocks down the worker out of the jobsite, and is permanent, even though the employee can return to usual work or another assignment after; e.g., loss of sight in one eye), permanent total (injury or illness that knocks down the worker out of the jobsite, and is permanent, and the employee may not return to usual work or another assignment after; e.g., wheelchair-bound injury), and death.
2) *Comprehensive or general liability insurance*: This third mandatory insurance covers third parties for bodily injuries to nonworkers or property damage associated with the project under construction and after. It protects the contractor and subcontractors in terms of third-party or nonworker lawsuits or claims from accidents or negligence associated with the project.
3) *Builder's risk or course of construction insurance (all-risk and named-perils)*: This is a type of property insurance that is mandatory and covers damage

to the project under construction by the contractor, as well as materials and equipment on site, from fire, storm, or water damage, explosion, vandalism, and theft as basic considerations.

4) *Other property insurance on projects under construction besides builder's risk insurance*: Other property insurance on projects under construction besides builder's risk insurance includes earthquake insurance, bridge insurance, steam boiler and machinery insurance, and installation floater insurance covering project machinery, such as heating, ventilating, and air-conditioning (HVAC) equipment from point of shipment to points of installation and testing on the project.
5) *Property insurance on project contractor's own property:* This type includes property insurance on the contractor's business premises or building, equipment or floater insurance covering the loss or damage of the equipment (regardless of where it is located) by schedule or blanket coverage, motor truck cargo policy, transportation floater insurance, fidelity bond or employee dishonesty insurance, crime insurance covering all perils, and valuable papers destruction insurance covering loss, damage, or destruction of valuable documents.
6) *Liability insurance or defense coverage:* This insurance encompasses coverage of the contractor against claims by third parties, and it includes:
   a) *General liability insurance:* Otherwise called public liability insurance or premises-operations insurance, it protects the contractor against third-party lawsuits and claims for bodily injury or property damage arising from its operations. With regard to the contractor's responsibility for liability to third parties' bodily injury, the concept of "attractive nuisance" applies where the contractor has a special obligation to protect children who may be attracted to trespass on the project under construction. Children have been shot for trespassing on the jobsite, and some have gone in and driven off dangerously with a tractor or a Bobcat.
   b) *Contingent liability insurance:* Otherwise called protective public and property damage liability insurance, it protects the contractor from its contingent liability under the law for third-party injuries or property damage arising from independent contractors (subcontractors and suppliers) that may not have had their insurance coverage. Hence, some insurers may request that the contractor obtain documentation attesting that the independent contractors have taken out the necessary liability insurance.
   c) *Personal injury liability insurance:* Personal injury is different from bodily injury, and refers to injury to reputation, defamation, slander, libel, false arrest, and invasion of privacy. Insurance can be taken out to cover all these potential occurrences.

d) *Professional liability insurance (PLI)*: Often referred to as professional indemnity insurance (PII), or errors and omissions (E&O) insurance, it protects construction project managers, architects, engineers, and other design professionals against lawsuits and claims arising from negligent errors and omissions in their work, e.g., legal fees and settlements resulting from judgment on the claims.
e) *Owner's protective liability insurance*: This insurance can be taken out by the contractor or the project owner, and it protects the owner from contingent liability for damages arising from the operations of the contractors and subcontractors.
f) *Workers' compensation insurance (workers' comp.)*: It is also a type of liability insurance—refer to its discussion earlier.
g) *Employer's liability insurance*: This insurance goes with workers' comp. and broadens its coverage. That is, it covers workers beyond workers' comp. coverage, or where workers' comp. is not available.
h) *Contractual liability insurance*: This insurance protects the contractor from the risk assumed on entering into a contract.
i) *Completed operations liability insurance*: This insurance protects the contractor from injuries to third parties or damage to property arising from completed projects.

7) *Employee insurance:* This includes employee benefits insurance for fringe benefits, unemployment insurance, disability insurance, and Social Security insurance.
8) *Automobile or collision insurance:* This insurance covers damage or loss of the contractor's vehicles, liability for bodily injury to others, or damage to property.
9) *Business, accident, and life insurance:* This includes:
   a) Business interruption insurance.
   b) Sole proprietorship insurance that helps heirs to continue or dispose of the business.
   c) Key person life insurance that covers the death and loss of the key person of the business.
   d) Corporate continuity insurance that helps the business to continue regardless of the loss of key personnel.
10) *Other insurance policies:* Other insurance policies that are obtained include:
   a) An umbrella excess liability insurance policy that provides extra coverage over commercial general liability, automobile, and employer's liability policies.
   b) Wrap-up insurance, which is a coordinated policy provided by the owner as holistic insurance as the project nears substantial completion.
   c) Nonoccupational disability insurance, which covers off-the-job injuries or illnesses.

Insurance is very essential in the construction process due to the many risks involved that can quickly imperil the contractor and the project. Consulting early with construction insurance agents or brokers is crucial to ensuring that all the necessary policies are in place before a project takes off. Insurance can be a trigger of disputes between the parties due to the following reasons:

1) *Extent of coverage*: Disputes may arise between the parties over the extent of coverage of an insurance policy; that is, what is covered or what is not included. Resources may be wasted in establishing or determining what is or is not covered.
2) *Responsibility for coverage and premiums*: Disputes may arise between the parties as to who is responsible for which insurance policy and/or its premiums. In other words, there may be ambiguity as to which party should take out a particular necessary policy and pay the premiums.
3) *Claim validity disputes*: Disputes may arise over the validity and/or extent of a claim. Denials or delays in claims settlement are a major potential source of disputes between the parties.
4) *Subcontractor default or underinsurance*: Disputes may arise if the subcontractors fail to maintain the necessary insurance policies, in which case the contractor takes responsibility for the liabilities and may cut corners or seek ways to recover from the owner.
5) *Differing interpretations of policies*: It is possible for the parties to have differing interpretations of the policy due to the specialized language in which most are written. Hence, disputes may arise over policy provisions, exclusions, and limitations.
6) *Coordination of policies*: There may be many policies held for one project, and disputes may arise during a claim as to which policy is primary and which is secondary, leading to uncertainties and delays in resolving and settling liabilities.

In order to prevent, avoid, or defuse disputes from insurance, it is essential for proactive risk assessments to be conducted prior to the project bid and after. The parties should clearly define and agree on the required insurance policies, the coverages, and the procedures for claims and resolutions if and when applicable. The owner should request and obtain all certificates of insurance from the contractor, and the contractor should do the same with the subcontractors and suppliers. Where insurance policy issues are not preventable or avoidable, open communication must be maintained between the parties to swiftly defuse any possibilities of disputes that will take time and cost extra on the project.

Please refer to Table 14.1 for the recommended PDM's D-PAD concept applicable in this regard.

References and Suggested Reading

Picken, D. and Ilozor, B.D. (2003). Height and construction costs of buildings in Hong Kong. *Construction Management and Economics*, 21, pp. 107–111.

Picken, D.H. and Ilozor, B.D. (2014). The relationship between building height and construction costs. In *Design Economics for the Built Environment: Impact of Sustainability on Project Evaluation (2015)* by Robinson, H., Symonds, B., Gilbertson, B. and Ilozor, B.D. eds., 47–60. John Wiley & Sons.

Chapter Review Questions

1) A contractor's financial capacity has no impact on project disputes. True/False?
2) Progress delays can lead to liquidated and ascertained damage. True/False?
3) Underbidding is a common cause of disputes in contractor bidding. True/False?
4) Using union labor always guarantees higher productivity. True/False?
5) Poor work quality can lead to rework and disputes. True/False?
6) Misuse of progress payments can lead to disputes with subcontractors and suppliers. True/False?
7) Which of the following is a contractor-related dispute flashpoint?
   a) Addenda
   b) Scope creep
   c) Financial capacity
   d) Owner's expectations
8) What is the purpose of a performance bond?
   a) To guarantee payment to subcontractors
   b) To ensure the contractor completes the project
   c) To cover insurance claims
   d) To prevent scope creep
9) Which of the following can cause labor productivity fluctuations?
   a) Clear weather
   b) Skilled labor
   c) Poor scheduling
   d) Adequate training
10) Which dispute flashpoint is related to the contractor's ability to manage resources?
    a) Unrealistic scheduling
    b) Site conditions

c) Payment delays
d) Project turnover

11) Which of the following is NOT a contractor-related dispute flashpoint?
a) Bonds
b) Insurance
c) Addenda
d) Workers' fidelity

12) ________ refers to the contractor's ability to manage funds, labor, materials, and equipment.

13) ________ damages are pre-agreed penalties for delays in project completion.

14) A contractor's ________ can affect trust and reputation in the industry.

15) ________ is the process of verifying that systems like HVAC and electrical are functioning properly.

16) ________ refers to the legal claim a contractor can place on a property due to nonpayment.

17) Which of the following is a key factor in equipment productivity fluctuations?
a) Operator skill
b) Weather conditions
c) Maintenance schedule
d) All of the above

18) Which of the following best describes the impact of underbidding?
a) Reduced project scope
b) Increased contractor profits
c) Financial strain and disputes
d) Faster project delivery

19) Contractors with poor safety records may face higher ________ ratings.
a) Commissioning
b) Experience modification
c) Liquidated damage
d) Insurance

20) Which PDM concept is most applicable when a contractor identifies a potential dispute early and takes steps to avoid it?
a) Dispute Prevention
b) Dispute Dousing
c) Dispute Resolution
d) Dispute Litigation

Exercises

1) **Equipment Productivity Fluctuations**

Scenario:

A contractor is building a new hospital wing. The project relies heavily on specialized excavation equipment. Midway through the project, several pieces of equipment broke down due to poor maintenance and operator error. Replacement parts are delayed, and the project falls behind schedule. The owner is frustrated by the delays and threatens to impose liquidated damages. The contractor argues that the delays were unavoidable and disputes the penalties.

Exercise 3.1 Applying D-PAD to Equipment Productivity Disputes

Instructions:

1) **Identify the Flashpoints:**
   - What specific contractor-related issues contributed to the dispute?
   - How did equipment maintenance and operator training factor into the problem?
2) **D-PAD Application:**
   - For each stage (prevention, avoidance, dousing), list at least two actions that could have been taken to reduce the risk or impact of this dispute.
   - Consider both technical (maintenance, scheduling) and communication (owner updates, documentation) strategies.
3) **Reflection:**
   - Which D-PAD stage would have been most effective in this scenario?
   - How could better planning or communication have changed the project outcome?

Deliverable:

Prepare a brief written summary or group discussion notes outlining your analysis and recommendations.

2) **Use of Union versus Nonunion Labor**
   Scenario:
   A contractor wins a large municipal project and plans to use a mix of union and nonunion labor to control costs. Early in the project, union workers stage a walkout, claiming that nonunion workers are being assigned tasks reserved for union members under the project labor agreement. Work stops for several days, and the project falls behind schedule. The owner is concerned about further disruptions and demands a resolution.

Exercise 3.2 Navigating Labor Disputes with PDM

Instructions:

1) **Analyze the Dispute:**
   - What were the root causes of the labor dispute?
   - How did the contractor's labor planning and communication contribute to the problem?

2) **D-PAD Strategy Development:**
   - Propose specific actions for each D-PAD stage (prevention, avoidance, dousing) that could have minimized or resolved the dispute.
   - Include strategies for contract review, stakeholder engagement, and on-site conflict resolution.
3) **Roleplay or Written Plan:**
   - In a group, roleplay a meeting between the contractor, union representatives, and the owner to resolve the dispute using your D-PAD strategies.
   - Alternatively, write a step-by-step action plan for the contractor to restore project momentum and prevent future labor disruptions.

Deliverable:
Submit a summary of your roleplay outcomes or your written action plan.

Discussions

1) **Underbidding and Accuracy of Estimates**
   Scenario:
   A midsized contractor, eager to win a competitive bid for a municipal library, submits a price significantly below the market average. The estimate was based on outdated material costs and optimistic labor productivity assumptions. Midway through construction, material prices spike and labor productivity falls short. The contractor faces financial strain, begins cutting corners, and requests multiple change orders. The owner feels misled and disputes the additional costs, leading to project delays and strained relationships. The contractor faces financial strain, begins cutting corners, and requests multiple change orders. The owner feels misled and disputes the additional costs, leading to project delays and strained relationships.
   How PDM Could Have Changed the Outcome:
   Dispute Prevention:
   - **Thorough Prebid Analysis:** The contractor should have conducted a comprehensive risk assessment, including up-to-date market research and realistic productivity benchmarks.
   - **Transparent Communication:** During prebid meetings, both owner and contractor could openly discuss cost assumptions, potential risks, and contingency plans.
   - **Contract Clauses:** Include clear language about how unforeseen cost escalations will be handled.

   Dispute Avoidance:
   - **Regular Estimate Reviews:** Schedule periodic reviews of project costs and productivity with the owner, allowing for early identification of discrepancies.

- **Collaborative Value Engineering:** Work together to find cost-saving alternatives before overruns occur.

Dispute Dousing:

- **Swift Issue Resolution:** When overruns become apparent, immediately convene a joint meeting to review the facts, adjust the scope, and negotiate fair compensation.
- **Documentation:** Maintain transparent records of all changes and communications to support resolution.

2) **Work Quality—Omissions, Defects, Failures, Rework, and Punch List**

Scenario:

A contractor is building a multistory office building. To save time and costs, the site supervisor skips some quality assurance checks. Several defects (e.g., improper window installation, uneven flooring) are discovered during the final walkthrough. The owner refuses to release the final payment until all issues are corrected, but the contractor disputes the extent of the defects and the cost of rework. Tensions escalate, and the project is delayed.

How PDM Could Have Changed the Outcome:

Dispute Prevention:

- **Robust QA/QC Program:** Implement a formal quality management plan with regular inspections, mockups, and preinstallation meetings.
- **Clear Specifications:** Ensure contract documents precisely define quality standards and acceptance criteria.

Dispute Avoidance:

- **Continuous Communication:** Maintain open dialogue with the owner and design team about quality expectations and progress.
- **Early Issue Identification:** Use punch list walk-throughs at intermediate stages, not just at project closeout.

Dispute Dousing:

- **Collaborative Problem-solving:** When defects are found, organize a joint inspection with all stakeholders to agree on the scope of rework.
- **Timely Remediation:** Prioritize quick correction of issues, document actions taken, and communicate progress to the owner to rebuild trust.

4

Subcontractor and Supplier-Related Dispute Flashpoints

Readers' Learning Objectives and Outcomes

1) *Beyond contract-related aspects, analyze additional subcontractor and supplier-related flashpoints and triggers of conflicts and disputes in the industry.*
2) *Conceptualize plausible prevention, avoidance, or dousing approaches that may be applicable to these conflicts' and disputes' flashpoints and triggers.*
3) *Match and apply prevention, avoidance, or dousing strategies effective or suitable to each conflict's and dispute's flashpoints and triggers.*

4.1 Introduction

Other than the privity of contract being between the general contractor (GC) and the subcontractor, the dispute flashpoints are closely similar to those between the owner and the general contractor, except in a few circumstances, such as underbidding and equitable estoppel, as well as conflicting work schedules and coordination. In other words, all the dispute flashpoints discussed above with regard to the privity of contract between the owner and the general contractor also apply to the privity of contract between the general contractor and the subcontractor or supplier, with the exception of the circumstances listed above. Similarly, the dispute flashpoints that apply in the privity of contract between the general contractor and the subcontractor also apply to the privity of contract between the subcontractor and the sub-subcontractor. However, due to the mechanics lien law limiting it to only second-tier relationships, we are not concerned here with the privity of contract between the subcontractor and the sub-subcontractor.

By subcontractors or suppliers, we mean those who have been recruited by the general contractor as specialty trades to take responsibility for carrying out certain tasks of the project, such as electrical, plumbing, and masonry. These trades play a collective, crucial role in the effective and efficient completion of projects. Here are the unique attributes and roles subcontractors bring to the project process:

1) *Provision of specialized skilled labor*: Subcontractors are engaged in the construction project process because they bring specialized skills needed to accomplish the project, such as carpentry, masonry, electrical, plumbing, heating, ventilating, and air-conditioning (HVAC). These tasks call for specialized knowledge, training, and skills that the general contractor may not have. Subcontractors have workers with the requisite skills for the specialty tasks, making it unnecessary for the GC to maintain a full-time workforce in all the specialty areas.
2) *Provision of specialized equipment*: The different specialized tasks require various equipment and experience in their use, which the contractor may not have.
3) *Certifications, licenses, and continuing education*: The different specialized tasks require certain certifications, licenses, and continuing education, which the contractor may not have.
4) *Work coordination*: Subcontractors coordinate these specialist tasks with the effectiveness and efficiency that contractors are not cut out for.
5) *Quality control*: The specialty contractors or subcontractors understand the interpretation of quality and control in their specialist task areas far better than the GCs.

6) *Quick, effective, and cost-efficient performance of work*: Due to their knowledge, training, skills, and experience in the specialist tasks, subcontractors are able to contribute to the quick, effective, and cost-efficient performance of the work far better than the GC doing them all alone. That's why in most nonspecialized projects, different from industrial and engineering projects, the GC's subcontracting to subcontractors can be as high as 75–85% of the work.
7) *Risk management*: Specialty contractors or subcontractors are better than GCs at assessing and managing the risks of their trades through insurance and other due diligence measures.

Although subcontractors and suppliers are as essential in the project process as their GC counterparts, both are potent sources of similar disputes (stemming from quality of work, delays, scope creep, cost overruns, payment issues, communication issues, liability, etc.), but the former additionally bring two distinct dispute aspects, namely: underbidding and equitable estoppel, as well as conflicting work schedules and coordination. The subcontractors, while offering quick and cost-efficient performance of work, also present problematic fragmentation of responsibility in the project process.

4.1.1 Underbidding and Equitable Estoppel

Underbidding is a fairly frequent occurrence among subcontractors, which is why they are often held to the law of equitable estoppel by bonds, as contractors rely on their numbers for the bids. If the contractor wins the bid, the subcontractors, by law, should stand by their numbers in utmost good faith, otherwise referred to as "uberrimae fidei." They should not estop on or back down from their numbers and leave the contractor to pick up the consequential extra costs. Underbidding happens due to several reasons that include:

1) *Rushed calculation*: Oftentimes, the subcontractors are given little time and tight schedules to do their estimates and submit their numbers. This can lead them to resort to contingencies and weird estimations, ending in miscalculations.
2) *Desperation for work*: A subcontractor may be motivated to underbid where there is a scarcity of work so as to enhance the chance to earn the job. They underbid to beat their competitors.
3) *Inexperience*: Inexperience can be a contributor to estimating inaccurately and underbidding. They may not properly assess the scope, complexity, and requirements of the work, and therefore unintentionally underbid.
4) *Underbidding with a remote agenda*: Subcontractors may deliberately underbid just to win the job, but strategically devise to recoup amounts from change orders and additional work down the line.

The legal principle of equitable estoppel is the protection of the general contractors from the underbidding reality of subcontractors. Underbidding is a big problem for both the GC and the subcontractors and can send the latter into bankruptcy if they have to bear the extra costs of their shortfalls. Underbidding and arguments over the application of the legal doctrine of equitable estoppel can be major sources of disputes between the parties that affect the time, cost, and quality of the project outcomes, and here are the possible scenarios:

1) *Underbidding*: Disputes may arise between the parties over extra costs, delays, and quality of work where the subcontractor underbid on the project. The GC will argue against bearing the extra costs and expect the subcontractor to complete the specialty task nonetheless. The subcontractor may argue over unfair pressure to cut corners to provide services for unaccounted or uncomputed tasks.
2) *Equitable estoppel*: Disputes arise between the parties when the GC invokes the doctrine of equitable estoppel to force the subcontractors to stick with their bid amount, upon which it relied to win the bid and award the job to them. The subcontractors may present indefensible justification for their bid numbers, which are unfortunately underbids. Such defenses may not be acceptable to the GC, and the law of equitable estoppel protects it and compels the subcontractors to abide by their numbers regardless that they are underbids.

In order to prevent, avoid, or douse disputes from subcontractors' underbidding and argument over equitable estoppel, the GC should allow ample time for the subcontractors to properly assess and price the job and provide reliable estimates they can stand behind. On the side of the subcontractors, they need to have the right training, skills, and experience in their specialty tasks, both in estimating and technical perspectives. They should avoid bidding under pressure, desperation, or with a strategic agenda to later recover shortfalls from change orders and extra work. If disputes nonetheless arise between the parties, open and effective communication should be swiftly deployed to douse them before they escalate. Having the appropriate bond protection helps, especially when the terms and processes are clearly spelled out in the contract from the onset.

Please refer to Table 14.1 for the recommended PDM's D-PAD concept applicable in this regard.

4.1.2 Conflicting Work Schedules and Coordination

Conflicting work schedules and coordination of subcontractors' work are common challenges in construction projects, particularly on complex projects with multiple trades and subcontractors. Here are some of the issues:

1) *Conflicting work assignments and schedules*: Different subcontractors are responsible for their specialty tasks, such as plumbing, electrical, and painting,

and each operates in its own timeline and schedules. Conflicts arise where the tasks overlap or clash; for example:

a) One subcontractor's work may depend on the completion of another's task; for instance, cast in situ slab work needs to be poured, cured, and set before the framing and partitions contractor can commence its work.
b) Access to the work of one subcontractor may be limited or made impossible by another subcontractor's work; the cast in situ slab example still applies here.
c) Two or more subcontractors may need scaffolds for their jobs, and there may be conflicts over who is responsible for the scaffolds and in what sequence they are to use them.
d) Project timeline changes or delays by one subcontractor can cause an unwieldy ripple effect on others' timelines and progress.

2) *Coordination of subcontractors' work:* It is essential that the different trades work seamlessly and efficiently to accomplish the project tasks on time, within the budget, and to quality. Effective coordination here involves the following:
   a) Marshal out the order, time, and sequence in which the subcontractors should effectively accomplish their project tasks to avoid crippling overlaps and/or clashes. For example, electrical and plumbing people should not be scheduled or sequenced to come before framing, which should precede them and be in place before them.
   b) Develop a master schedule for all engaged subcontractors that proactively resolves work overlaps and clashes to eliminate or limit them to the barest minimum, where they don't delay the project delivery timeline or distort the milestones.
   c) Manage site logistics to ensure that each subcontractor has access to its materials, equipment, and workspace to effectively and efficiently perform their work tasks.
   d) Facilitate open and effective communications among the subcontractors and with the GC and other participants to share information, swiftly address issues, and coordinate activities.

Conflicting work schedules and coordination of subcontractors can be a major source of disputes in construction, and here are some issues that bring them about:

1) *Productivity and delay disputes from conflicting work assignments, overlaps, and/or clashes*: Electrical work may overlap or clash with plumbing, and if the sequence is not clearly worked out and communicated timely, there may be fights as to who is to go first, and one may delay the other's work. The subcontractors can blame one another or the GC for poor management that overlooked work-assignment overlaps and/or clashes. The delays can cascade to a project timeline extension with its consequential damages.

2) *Costs overrun disputes from the delays:* There may be disputes as to who is responsible for the liabilities of delays. The delayed subcontractors may seek compensation for their downtime and time extension from the GC, which may dispute such requests.
3) *Disputes over resources:* Subcontractors may fight over materials and equipment, especially when there is a shortage. Fighting over scaffolds is a ready example.
4) *Safety concerns:* Safety concerns increase when there are many people working their different trades at the same time. There will be different attitudes to issues of safety and procedures. When some accidents or incidents happen, work can be delayed if not halted altogether.
5) *Quality concerns:* Quality may be affected when there are too many people working their different trades at the same time. Some may damage others' work, leading to inevitable rework and extra costs of money and time.
6) *Territorial disputes:* This can be a problem with some subcontractors who may be averse to sharing their workspace amicably with others.

Preventing or avoiding these disputes requires preplanning, proactive coordination, and collaborative problem-solving among the subcontractors, facilitated by the GC. Experience and skill in appropriately scheduling and managing subcontractors in the particular project type are crucial in avoiding these disputes. Effective and open communication is essential in swiftly dousing any disputes seeking to arise here.

Please refer to Table 14.1 for the recommended PDM's D-PAD concept applicable in this regard.

Chapter Review Questions

1) Subcontractor underbidding is often driven by desperation or lack of experience. True/False?
2) Equitable estoppel protects subcontractors from being held to their bid amounts. True/False?
3) Conflicting work schedules among subcontractors rarely lead to disputes. True/False?
4) Subcontractors and suppliers have no influence on project delays or cost overruns. True/False?
5) Effective coordination among trades can help prevent disputes. True/False?
6) Subcontractors are typically responsible for 75–85% of the work in nonspecialized projects. True/False?
7) Which of the following is a unique dispute flashpoint for subcontractors?
   a) Payment delays
   b) Scope creep

c) Underbidding and equitable estoppel
d) Change orders

8) What is the primary legal principle that holds subcontractors to their bid amounts?
   a) Privity of contract
   b) Equitable estoppel
   c) Mechanics lien
   d) Liquidated damages
9) Which of the following is a recommended strategy to avoid coordination disputes?
   a) Ignore overlapping schedules
   b) Use pull planning and visual work area maps
   c) Assign all trades to the same area
   d) Delay subcontractor engagement
10) Which of the following is NOT a cause of subcontractor underbidding?
   a) Strategic planning
   b) Rushed estimation
   c) Inexperience
   d) Desperation for work
11) What is a common consequence of conflicting work schedules among subcontractors?
   a) Improved productivity
   b) Reduced costs
   c) Project delays and disputes
   d) Faster inspections
12) Poor coordination among subcontractors can result in delays, cost overruns, and compromised ________.
13) Subcontractors often provide specialized ________ and equipment for specific tasks.
14) Conflicting work schedules can lead to disputes over ________ and task sequencing.
15) Subcontractors may underbid due to tight bidding timelines or ________ for work.
16) Effective dispute management among subcontractors includes early planning and clear ________.
17) Which of the following is a best practice for managing schedule conflicts among trades?
   a) Assign overlapping tasks
   b) Avoid communication
   c) Hold daily huddles and weekly meetings
   d) Delay project start

18) Which of the following is a potential result of subcontractor underbidding?
    a) Increased profits
    b) Bankruptcy
    c) Reduced scope
    d) Faster delivery
19) Subcontractor disputes can be minimized by allowing ample time for accurate ________.
    a) Estimating
    b) Marketing
    c) Hiring
    d) Auditing
20) Which PDM concept is most applicable when a subcontractor dispute begins to emerge and needs to be quickly neutralized?
    a) Dispute Prevention
    b) Dispute Avoidance
    c) Dispute Dousing
    d) Dispute Resolution

Exercises

Exercise 4.1 Subcontractor Bid Evaluation and Risk Analysis

Objective:
Understand how underbidding and equitable estoppel can lead to disputes; apply PDM strategies to mitigate risks.

Instructions:

1) Select a real or hypothetical construction project involving multiple subcontractors.
2) Review sample bids submitted by subcontractors for specialized trades (e.g., electrical, plumbing).
3) Identify:
   - Any bids that appear significantly lower than others.
   - Potential risks associated with accepting the lowest bid.
4) Conduct a mock dispute-risk analysis using PDM principles:
   - What disputes could arise from underbidding?
   - How can equitable estoppel be applied?
   - Which D-PAD strategy (prevention, avoidance, dousing) would be most effective?

Outcome:
Gain insight into how bid evaluation affects dispute potential and how proactive strategies can protect project integrity.

Exercise 4.2 Coordination Simulation Among Subcontractors

Objective:
Explore how conflicting work schedules and poor coordination among subcontractors can lead to disputes; practice applying PDM concepts to resolve them.

Instructions:

1) *Form a group and assign roles*: General contractor, electrical subcontractor, plumbing subcontractor, and site supervisor.
2) Use a fictional scenario where overlapping tasks (e.g., electrical and plumbing installations) are scheduled in the same area at the same time.
3) Roleplay a coordination meeting to:
   - Identify the conflict.
   - Discuss responsibilities and sequencing.
   - Apply D-PAD strategies to prevent escalation.
4) Reflect on how better planning and communication could have avoided the conflict.

Outcome:
Develop practical skills in dispute avoidance and dousing through collaborative coordination and scheduling.

Discussion

1) How can general contractors proactively manage subcontractor underbidding to prevent disputes during project execution?
   - Consider the role of bid vetting, equitable estoppel, and communication during pre-construction.
   - Discuss how PDM principles (especially dispute prevention and avoidance) can be applied to reduce risk.
2) What strategies can be used to coordinate multiple subcontractors effectively and avoid scheduling conflicts on complex projects?
   - Explore real-world examples of coordination failures and their consequences.
   - Debate the effectiveness of tools like pull planning, daily huddles, and visual work area mapping in dispute dousing.

5

Construction Documents-Related Dispute Flashpoints

Readers' Learning Objectives and Outcomes

1) *Analyze the various flashpoints and triggers of conflicts and disputes in the industry stemming from contract documents.*
2) *Conceptualize plausible prevention, avoidance, or dousing approaches that may be applicable to these conflicts' and disputes' flashpoints and triggers.*
3) *Match and apply prevention, avoidance, or dousing strategies effective or suitable to each conflict's and dispute's flashpoints and triggers.*

5.1 Introduction

Construction documents (not contract documents) refer to the drawings, schedules, and other written, nongraphical information prepared by architects, engineers, and other designers to communicate the project intent as it must be built and delivered by builders or contractors. Project information that is better communicated or represented by graphics is provided as drawings, and that which is better conveyed in words is provided as specifications, schedules, and other written forms. Drawings and specifications are complementary to each other, and neither is sufficient to completely convey a building project's information. Pieces of information that are better presented graphically are drawn as drawings, while those that are better conveyed in verbal or written form accompany as specifications. They are read and interpreted in tandem as the project is being constructed. Construction documents vary, as they come in detail depending on the architects, engineers, and other designers who prepare them. Some may be overdetailed, under-detailed, or just right (neither overdetailed nor under-detailed). Construction documents are comprised of:

1) *Architectural drawings and specifications*: These are graphical representations of the project or structure as site plans, floor plans, roof plans, elevations or views, building and wall sections, details, schedules, and quantities, and notes. The drawings show the project situated and oriented on the site, its layout, dimensions, and annotations of spaces and components. Aspects of the structure that are better represented descriptively in words accompany the drawings as architect's specifications, or more expansively, as project manuals.
2) *Structural drawings and specifications*: The drawings here are prepared by structural engineers relative to the architect's design, showing and detailing the foundations, slabs, beams, columns, and load-bearing walls and their reinforcements, as well as the roof framing with the necessary schedules and quantities, as well as dimensions and notes. They specify the structural materials, sizes, and their placements. Just as with architectural drawings, the structure's information that is better presented graphically is drawn as a drawing, while those that are better conveyed in verbal or written form accompany as specifications.
3) *Mechanical, electrical, and plumbing (MEP) drawings and specifications*: The drawings are prepared by electrical and mechanical engineers to accompany architectural and structural drawings showing and detailing heating, ventilating, and air-conditioning (HVAC) systems, electrical power and fixture installations, plumbing and firefighting systems, with the necessary schedules and

quantities, as well as dimensions and notes. They specify the materials, sizes, and their placements. Just as with architectural and structural drawings, the MEP information that is better presented graphically is drawn as a drawing, while those that are better conveyed in verbal or written form accompany as specifications.

4) *Civil engineering drawings and specifications*: Civil engineers deal with the drawings and specifications here. The drawings show and detail all the horizontal groundwork and features of the project, like the driveways, drainage, services, and utilities, with the necessary schedules and quantities, as well as dimensions and notes.
5) *Landscape drawings and specifications*: Landscape architects or horticulturists deal with the drawings and specifications here. The drawings show the changed and retained land features, such as trees, turf, site furniture, and mounds, with their placements and the necessary schedules and quantities, and notes.
6) *Survey drawings and specifications*: Land surveyors deal with the drawings and specifications here. The drawings show and detail the property boundary, site topography and features, with the necessary schedules and quantities, as well as dimensions and notes. They situate the site on its cardinal location, showing the North, East, West, and South cardinal directions.
7) *Demolition drawings and specifications*: The drawings can be prepared by architects and other designers where there are existing structures that may be partly or wholly demolished. They show what is to be demolished and/or retained, as well as dimensions and notes.
8) *Interior drawings and specifications*: The drawings are prepared by interior designers, and they show all interior fixed and movable features, furniture and furnishings, as well as decorations, with the necessary schedules and quantities, as well as dimensions and notes.
9) *Communication drawings and specifications*: Commonly applicable to larger and more complex or sophisticated projects, the drawings are prepared by specialist communication engineers showing and detailing communication-related installations and network, security, and surveillance systems.
10) *Technology and/or process or factory drawings and specifications*: Commonly applicable to larger and more complex projects, the drawings are prepared by specialist engineers and designers showing more technical and technological installations not accommodated in conventional architectural and MEP drawings. Examples are commercial kitchens and projects with process or factory-like installations.
11) *Priced and unpriced bill of quantities*: These are neither classified as drawings or specifications but are part of the construction document that are prepared

by quantity surveyors and cost engineers detailing all the project components with and without their costs. This aspect of the documents is useful for the bidding process and for controlling and monitoring the project throughout its construction to ensure that it is working with and within the budget.

Construction documents play a central role in the successful delivery of the project for all participants and stakeholders, ensuring that it is built to the requirements and standards as intended and designed. References are made to them throughout the process and are the major tool in benchmarking and measuring quality attainments. Nonetheless, due to their characteristics and variations, construction documents can be a source and a trigger of disputes among the parties in the following broad aspects:

5.2 Quality of Drawing Documents

The quality of drawing documents is a measure of their effectiveness in clearly, completely, and graphically conveying the design intents of the project to all involved in building it according to the requirements and standards desired. Here is a breakdown of the attributes constituting the quality of drawing documents:

1) *Accuracy*: Drawing documents must correctly represent the design intents of the architects, engineers, and/or other designers in their interpretations and translations of the owner's desires and programs. They should embody correct location or siting, layout, precise dimensions and details, and identification and notation of materials and components. Misspellings and typos can be expensive in notes and annotations, as in specifications. There was a case of a mistaken representation of "integral" as "internal" in a glazing mullion label and specification in the past that cost a lot of time and money to correct.
2) *Clarity*: This attribute ensures that the design information intended to be conveyed to all project participants and stakeholders (owners, contractors, subcontractors, etc.) is unambiguous, easily and equally understood by everyone without conflicting readings and interpretations. Legibility is key, as is the inclusion of symbols, labels, and annotations.
3) *Completeness*: The attribute calls for the drawings to be comprehensive in covering all and every aspect of the project, from complete architectural, structural, and MEP drawings to all other necessary drawings and details, as well as their dimensions and notes. Any missing aspect affects the project through confusion, requests for information (RFI), change orders, and consequential delays and costs.
4) *Consistency*: The drawings must be consistent and coordinated across all series—architectural, structural, MEP, etc. The drawing series must align and

be concordant with one another as collectively speaking to the same project. Series A drawings are architectural drawings, C are civil, S are structural, M are mechanical, E are electrical, P are plumbing, and T are technology.

5) *Clutter-free:* Drawings must be clutter-free by not trying to infuse most or all specification stuff on the drawings. Cluttered drawings increase the chances of confusion and research time. Pieces of information that are better presented graphically are drawn as drawings, while those that are better conveyed in verbal or written form accompany as specifications.
6) *Detail:* Drawings must be sufficiently detailed to comprehensively convey design intent by material, placement, proportion, and measurement. Details convey information about construction methods, material specifications, finishes, instructions, and other vital pieces of information that complete the message transmitted.
7) *Codes and standards*: Drawings should completely comply with all applicable codes, standards, and regulations for the project from the controlling authorities and bodies. This includes safety standards, sustainability requirements, and fire codes.
8) *Currency and accessibility*: Drawings must be prepared with current and accessible tools and technologies for all project stakeholders, whether they are hard copies, portable document format (PDF), computer-aided design and drafting (CADD), building information modeling (BIM), or other formats. Compatibility of the drawings facilitates document accessibility, sharing, and collaboration among the project participants and stakeholders.
9) *Update*: Drawings should be updatable, and such is captured through a record of revisions on the drawings. This helps in monitoring and documenting all changes, and eventually preparing as-built drawings is helpful in the operations and maintenance management of the structure through its lifecycle.

Essentially, quality drawing documents facilitate effective communication, coordination, and elimination or minimization of errors, omissions, and commissions, as well as delays and costs in the project delivery process. They are the primary means of conveying the project design intent, and with other accompaniments like specifications, schedules, and notes, constitute complete and comprehensive information for the successful delivery of the project on time, within budget, and to quality.

However, if the quality of drawing documents is in question, it can bring about disputes in the following ways:

1) *Errors, omissions, or commissions*: Negligent drawing documents replete with errors, omissions, or commissions are potent grounds for confusion, conflicts, and disputes due to their impact on time, cost, and quality that are the express and implied contract terms. Errors come in the form of misrepresentation or

wrong representations. Omissions mean that vital aspects are oversighted or not considered or included, and commissions mean including what should not be included that ultimately misleads and brings about confusion and deviations from the project design intent.

2) *Misinterpretations*: If the drawings lack clarity and precision, they are prone to misinterpretation or varying interpretations, of which correction may consume the resources of time, materials, labor, equipment, and money. Disputes can arise as a result.
3) *Discrepancies*: The drawings may be discrepant with the specifications, and these are breeding grounds of incorrect construction that would be costly to correct or rework in terms of the resources of time, materials, labor, equipment, and money. Coordination is difficult under this scenario, and disputes can ultimately arise as a result.
4) *RFI and change orders*: Poor-quality drawing documents present confusion, challenges, and call for more RFIs and change orders; all of that can introduce additional complexity and uncertainty as the project progresses, thus altering the scope as well as the intents and objectives of the project in terms of time, cost, and quality. This can bring about disputes between the parties.
5) *Delays from regulatory authorities*: Poor-quality drawing documents that fail to consider and accommodate regulatory requirements, codes, and standards attract negative regulatory oversights that can cause delays or ultimately halt the project altogether. This can cause disputes between the parties if the project timeline is affected, with the consequential damages and the fights over who is to take the liability and responsibility.
6) *Breach of contract*: As part of the contract documents, poor-quality drawing documents can facilitate a breach of contract in many ways. Omitting any aspect or wrongly building any aspect as a result of poor-quality drawing documents constitutes a breach of contract that is a ready trigger of disputes between the parties.

In order to prevent or avoid disputes from poor-quality drawing documents, they should not be used to break ground on the project to begin with. The drawings must be made to be of sufficient quality to successfully deliver the project's design intents and objectives on time, within the budget, and to quality. Quality assurance (QA), quality control (QC), total quality management (TQM), Six Sigma, and/or ultimately PDM principles and precepts should be embraced to ensure that the drawing documents are complete and comprehensive to communicate precisely and effectively to lead the way to a successful project outcome. Where any drawing issues arise, nonetheless, swift action should be taken to respond to them to avoid escalation to disputes.

Please refer to Table 14.1 for the recommended PDM's D-PAD concept applicable in this regard.

5.3 Quality of Specifications

The quality of specifications is a measure of their effectiveness in clearly and completely conveying the design intents of the project to all involved in building it in written form according to the requirements and standards desired. Quality specifications should supplement the drawing documents rather than duplicate them. The attributes constituting the quality of specifications are quite similar to those of the drawings, with a couple more as follows:

1) *Verbiage*: Specification writing requires a special tact and skill. A quality specification is precise and succinct, easily distinguishable by type that includes prescriptive (otherwise called design or material and workmanship specification—open and closed types, specifies in prescriptive terms the types of materials and manufacturers to use and the specific design and performance applications), performance (specification that defines or specifies in descriptive terms the performance or service requirements and expected outcomes rather than the technical requirements of the project components), and technical (specification that stipulates the material and workmanship standards requirements of the project; e.g., American Society for Testing and Materials [ASTM], American National Standards Institute [ANSI]).
2) *Standalone*: Quality specifications must be complete and able to stand alone without needing graphics to explain them.

Disputes caused by poor specifications are similar to those of poor-quality drawings, and the fixes are exactly the same.

Please refer to Table 14.1 for the recommended PDM's D-PAD concept applicable in this regard.

5.4 Discrepancies

Discrepancies in construction occur when there are ambiguities or disparities within graphical or written information in the documents—drawings, notes, specifications, and written instructions. The design intent is confused or conflicted as a result. Discrepancies manifest in various scenarios that include:

1) *Drawing dimensions versus specifications*: There may be variations between the measurements on the drawings and the design specifications.
2) *Alternate disparity*: There may be a disparity between allowed alternates as to their equality in quality and performance.
3) *Quality disparity*: Beyond alternate disparity, quality disparity can also apply to workmanship, which can mean a difference between the quality delivered and that expected.

4) *Change order disparity*: A change order can bring about a distinction between the original design intent and the change ordered. This disparity can extend to layout, scope, aesthetics, and quality.
5) *Codes and standards disparity*: Any aspect of the project not built to codes and standards is discrepant with the applicable codes and standards.
6) *Schedule discrepancies*: Delays that change the project schedule are discrepant with the contractual delivery timeline.
7) *Cost discrepancies*: Cost overruns that change the project's cost are discrepant from the contractual construction cost.
8) *Safety discrepancies*: Where safety protocols and requirements are not followed, they are discrepant from the prescribed safety standards and compliance.

Discrepancies are big issues in the construction process, and if not managed proactively and effectively, could result in chaotic and disputatious project delivery in the following ways:

1) *Design errors, omissions, and commissions disputes*: Discrepancies largely emanate from these sources and can be major causes of disputes.
2) *Fulfilling contractual obligations*: Discrepancies can be a clog in the wheel of contractual obligation fulfillment with regard to agreed-upon design drawings, specifications, schedules, and addenda.
3) *Disputes from change orders*: Change orders can bring about discrepancies in the original agreed-upon terms, scope, schedule, and costs to the extent that they lead to disputes between the parties.
4) *Quality of materials and workmanship disputes*: Disputes can arise over the equality in quality and performance of alternate materials as much as they can arise over actual versus expected workmanship quality.
5) *Discrepant cost disputes*: Cost disputes arise over the discrepancy between the agreed-upon project budget and the actual costs affected by overruns.
6) *Discrepant schedule disputes*: Schedule disputes arise over the discrepancy between the agreed-upon project timeline and the actual timeline as extended by delays.
7) *Scope creep disputes*: Scope creep disputes arise over the discrepancy between the agreed-upon project scope and the actual final scope as enlarged by changes and site conditions.
8) *Regulatory compliance disputes*: These disputes arise over the discrepancy between prescribed standards (codes, safety, environmental, etc.) and noncompliant project aspects.

In order to prevent or avoid disputes arising from discrepancies, it is essential to abide by the culture of working with comprehensive and thorough documentations vetted for QA/QC/TQM, Six Sigma, and PDM. Clear, effective, and swift communication is helpful in dousing any potential disputes before they escalate to needing resource-consuming negotiation, mediation, arbitration, or litigation.

Please refer to Table 14.1 for the recommended PDM's D-PAD concept applicable in this regard.

5.5 Reading and Interpretation of Construction Documents

Construction documents are the primary means of communication among project participants and stakeholders for successful delivery. The ability to read and interpret construction documents is crucial in the project delivery process. The documents are prepared by different architects, engineers, and designers, and can vary as they come in details, accuracy, legibility, and complexity. The reading and interpretation of construction documents involves:

1) *Understanding document types*: Document types include architectural, engineering, drawings, specifications, schedules, and notes. These documents are interconnected and must be read and interpreted as such. A mastery of the symbols and acronyms is important to accurately reading and interpreting the documents.
2) *Communicating and coordinating document implementations*: The ability to accurately and effectively translate and interpret the documents in the project process requires skills and experience.
3) *Request for information and associated change*: Capable reading and interpretation of construction documents enables timely identification of discrepancies, errors, omissions, and commissions in them and an appropriate request for information that may result in changes.

Misreading and misinterpreting construction documents can be a major cause of disputes among the parties, and here are how:

1) *Disputable ambiguities and discrepancies*: Misread and misinterpreted construction documents can bring about disputatious ambiguities and discrepancies between what is built and what was designed. With ambiguities and discrepancies, the documents can be differently read, understood, and interpreted by various project participants and stakeholders, and these conflicting takes and misunderstandings are a major cause of disputes.
2) *Disputable errors, changes, revisions, and rework*: Misread and misinterpreted construction documents can lead to disputatious errors, work changes, revisions, and rework.
3) *Breach of contract*: Misread and misinterpreted construction documents can lead to a breach of contract by delivering a project that is not in compliance with the agreed-upon contract.

4) *Scope creep*: Misread and misinterpreted construction documents leading to scope creep can trigger disputes between the parties over who is to bear the liabilities and responsibilities of the increased scope.
5) *Schedule overrun*: Misread and misinterpreted construction documents leading to schedule overruns occasioned by delays, revisions, corrections, and rework can trigger disputes between the parties over who is to bear the associated damages of an extended project delivery timeline.
6) *Cost overrun*: Misread and misinterpreted construction documents leading to cost overruns occasioned by changes, revisions, and rework can trigger disputes between the parties over who is to bear the additional costs beyond the contractual construction cost or budget.
7) *Regulatory compliance disputes*: Misread and misinterpreted construction documents leading to noncompliance with codes, standards, environmental, and safety requirements can trigger disputes between the parties over the responsibility and liability for noncompliance.
8) *Quality and performance standards*: Misread and misinterpreted construction documents may come in the form of discrepant material and workmanship quality and standards, and such can trigger disputes between the parties over the delivered and designed or agreed-upon standards and performance.

In order to prevent or avoid disputes from misread and misinterpreted construction documents, there must exist a culture of thorough document reading and interpretation anchored on the principles of QA/QC/TQM, Six Sigma, and PDM. There must be a recognition that there are differences in the details, accuracy, legibility, and complexity of documents prepared and provided by different architects, engineers, and designers, and Proactive efforts must be made to analyze and assess the risks, clarify, and thoroughly comprehend the documents ahead of breaking ground for the project. Whenever aspects are not captured or fail to show up before project commencement, clear, effective, and swift communication must be deployed to douse any potential disputes before they escalate to need resource-consuming negotiation, mediation, arbitration, or litigation. It is therefore imperative to promote thorough preconstruction and in-construction document review, clear communication, and proactive resolutions of complexities, discrepancies, and ambiguities.

Please refer to Table 14.1 for the recommended PDM's D-PAD concept applicable in this regard.

Chapter Review Questions

1) Errors and omissions in construction documents can lead to scope creep and disputes. True/False?

2) Specifications are always sufficient on their own to convey complete project intent. True/False?
3) Discrepancies between drawings and specifications are a common source of disputes. True/False?
4) Clear communication is not necessary when interpreting construction documents. True/False?
5) Poor detailing in drawings can result in delays and cost overruns. True/False?
6) Reading and interpreting construction documents is a skill that can help prevent disputes. True/False?
7) Which of the following is a major flashpoint in construction documents?
   a) Weather conditions
   b) Quality of drawing documents
   c) Labor union disputes
   d) Site access
8) What is the best way to prevent disputes from discrepancies in construction documents?
   a) Ignore minor errors
   b) Use verbal agreements
   c) Conduct thorough document reviews
   d) Delay construction
9) Which of the following is NOT typically included in construction documents?
   a) Drawings
   b) Specifications
   c) Verbal instructions
   d) Schedules
10) What does poor coordination between drawings and specifications often lead to?
   a) Faster project delivery
   b) Disputes and rework
   c) Reduced costs
   d) Improved quality
11) Which PDM concept is most applicable when discrepancies are identified early and addressed before construction begins?
   a) Dispute resolution
   b) Dispute litigation
   c) Dispute prevention
   d) Dispute arbitration
12) ________ refers to inconsistencies between drawings, specifications, and other project documents.
13) Construction documents must be reviewed for errors, omissions, and ________ before construction begins.

14) The quality of ________ documents directly affects how well a project is understood and executed.
15) ________ breakdown is a common trigger of disputes when interpreting construction documents.
16) Scope ________ occurs when the project expands beyond its original intent due to unclear documentation.
17) Which of the following is a recommended strategy to avoid disputes from poor specifications?
    a) Use generic descriptions
    b) Rely on contractor judgment
    c) Write clear, detailed specifications
    d) Skip specification review
18) Which of the following is a consequence of poor document handling?
    a) Increased collaboration
    b) Reduced project scope
    c) Delays and cost overruns
    d) Improved safety
19) Construction documents should be reviewed by all internal project team members before ________.
    a) Final payment
    b) Bid closing
    c) Site inspection
    d) Warranty issuance
20) Which of the following is a key benefit of applying PDM to construction document disputes?
    a) Increased litigation
    b) Reduced communication
    c) Hitch-free project outcomes
    d) Higher insurance premiums

Exercises

Exercise 5.1 Document Discrepancy Audit Simulation

Objective:
Identify and resolve discrepancies between drawings and specifications using PDM strategies.

Instructions:
Choose a real or hypothetical construction project with a full set of drawings and specifications.

Review the documents and identify at least three discrepancies (e.g., conflicting dimensions, material mismatches).

For each discrepancy:
Describe the potential dispute it could cause.
Apply dispute prevention, avoidance, and dousing (D-PAD) strategies to resolve or mitigate the issue.
Present your findings in a short report or group presentation.

Outcome:
Develop skills in proactive document review and dispute prevention through early detection and resolution.

Exercise 5.2 Miscommunication Roleplay Based on Document Interpretation

Objective:
Explore how misinterpretation of construction documents can lead to disputes and practice dousing techniques.

Instructions:
Form a group and assign roles: Architect, contractor, owner, and site supervisor.
Use a fictional scenario where a note on a drawing is interpreted differently by each party (e.g., "or equal" clause in material specification).

Roleplay a meeting to:
Discuss each party's interpretation.
Identify the root cause of the miscommunication.
Apply PDM dousing strategies to reach a resolution.
Reflect on how clearer documentation or communication could have prevented the issue.

Outcome:
Gain practical experience in resolving disputes through collaborative communication and clarification.

Discussion

1) How can construction teams ensure that drawings and specifications are coordinated to prevent disputes?
   - Discuss the role of document control, QA/QC processes, and collaborative reviews.
   - Explore how PDM principles can be applied during preconstruction to catch inconsistencies early.

2) What are the risks of relying solely on specifications without cross-referencing drawings, and how can those risks be mitigated?
   - Consider examples of misinterpretation or omissions.
   - Debate the importance of training in document reading and interpretation for dispute prevention.

6

Contract-Related Dispute Flashpoints

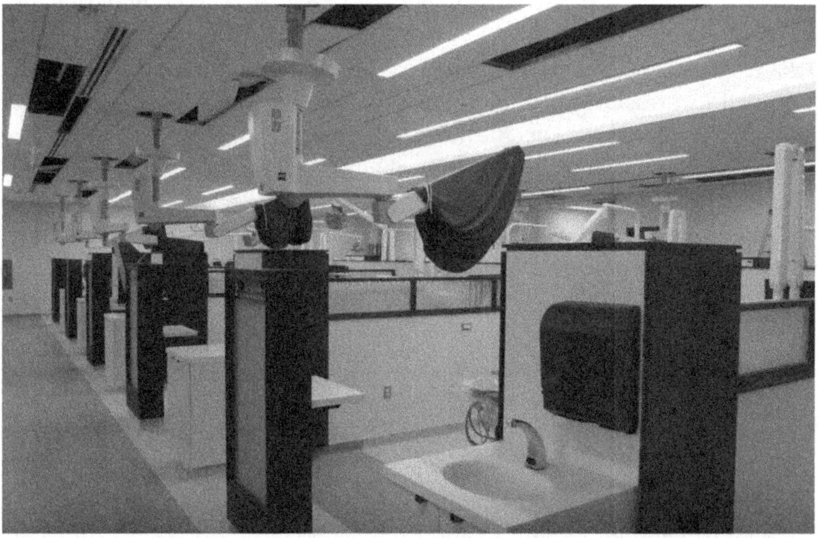

Readers' Learning Objectives and Outcomes

1) *Analyze the various flashpoints and triggers of conflicts and disputes in the industry beyond project owners and contractors.*
2) *Conceptualize plausible prevention, avoidance, or dousing approaches that may be applicable to these conflicts' and disputes' flashpoints and triggers.*
3) *Match and apply prevention, avoidance, or dousing strategies effective or suitable to each conflict's and dispute's flashpoints and triggers.*

6.1 Introduction

A contract in construction is a legally enforceable written agreement between two parties (for instance, between a project owner and a project contractor) outlining and guiding the content, terms, and conditions of project delivery from start to finish. The key aspects of a construction contract include:

1) *Parties to contract or privity of contract*: The privity of contract is between two parties—for instance, between a project owner as the contract offeree and the project contractor as the contract offeror. In subcontracts, the privity of contract is between the general contractor as the offeree and the subcontractor or supplier as the offeror. There are Additional parties that can enter into a contract or an agreement with the owner in the construction process, such as architects, construction managers, engineers, or other designers and consultants.
2) *Project scope*: The contract defines and details the project scope in what are termed "contract documents" that include the contract or agreement between the parties, drawings, specifications, bid documents, general conditions, special conditions, schedules, addenda, change orders, and other documents that explain or outline what needs to be done in the project.
3) *Contract sum and payment terms and schedules*: The contract specifies the contract sum and the schedules for its interim or progress payments as the project advances relative to some milestones or agreed timeframes and draws. It indicates any percentage retentions or withholding amounts by the owner from interim payments and when or how those retentions will be released to the contractor.
4) *Project timeline*: The contract specifies the start and delivery dates, including schedules and milestones in between. It stipulates how any issues of delays and time extensions will be handled.
5) *Codes, regulations, quality standards, and specifications*: The contract states the prevailing codes, regulations, standards, and specifications to adhere to in the project in terms of materials and components, workmanship, and performance.
6) *Obligations and responsibilities*: The contract delineates the obligations and responsibilities of the parties and participants in the project—owner, contractor, subcontractor, suppliers, and consultants. For instance, it is the obligation and responsibility of the project owner to give possession of the site to the contractor and guarantee that the site is fit to bear the project to be sited on it.
7) *Changes in the project*: The contract stipulates how changes will be handled, both those changing the scope, cost, and time of the project and those that

don't. It states the modalities for change orders, requests for information, approval, and implementation.

8) *Dispute handling*: The contract anticipates disputes and stipulates how to handle them when they show up. For instance, most contracts may require that any dispute first be mediated or arbitrated before litigation.
9) *Bonds, insurance, and indemnification*: The contract may specify the bonds and insurance policies to retain for the project and what indemnifications may apply for which party.
10) *Termination*: The contract states the conditions under which it may be terminated between the parties and the consequences of such termination to either party.

Generally, contracts are among the most essential aspects of construction projects, but they can be potent sources of disputes from ambiguities in the contract documents, discordant understanding and interpretation of the clauses, omissions and commissions, and sharing of risks and rewards in the contracts.

6.2 Ambiguities in the Contract

If the contract is not comprehensive and precise, has inconsistencies and discrepancies, it is said to be ambiguous and subject to multiple interpretations and misinterpretations. The language of the provisions of the contract may be imprecise, unclear, vague, ambivalent, and prone to multiple interpretations and misinterpretations. What is conveyed by the drawings may be inconsistent with the written specifications. Even typos in the written documents can skew the interpretation, as when "integral" was mistakenly written as "internal" in glazing bars. Here are some common ambiguities that can be found in contract documents:

1) *Undefined terms in written documents*: Some terms are by nature ambiguous if not defined when used. They may have multiple meanings or be open to multiple interpretations. For instance, in addenda or alternates, the use of "or equal" has led to arguments over the establishment of equality between alternate products. The use of the words "reasonable" and "adequate" can attract dissension if used but not defined precisely.
2) *Contradictory or conflicting clauses and provisions*: Sections and/or clauses of the contract may contradict or conflict with another, or the aspects of the specifications may contradict or conflict with the drawings. The standard contract form used may be incongruent or inconsistent with the particular project requirement and circumstances. Contradictions or conflicts often arise where the culture of quality assurance (QA), quality control (QC), total

quality management (TQM), Six Sigma, and PDM is nonexistent, and rampant negligence is commonplace.

3) *Sketchy details*: When the contract documents are not detailed sufficiently, they may not capture the entire scope of the project. Where details are lacking, the temptation to fill in the gaps is rife, and such can introduce ambiguities. There will be differing expectations between the parties under these circumstances.
4) *Change management, payment terms, and schedules*: When payment terms and schedules are not clearly defined, ambiguity can arise over changes and additional work, as well as when to pay and what to retain.

Ambiguities in contract documents are a major headache in the project construction process and can lead to disputes between the parties in the following ways:

1) *Different interpretations and expectations*: Ambiguities in the contract documents lead to different interpretations and expectations that show up in building the project wrongly or below the expected standards and requirements. Hence, disputes can arise over differing expectations and deliverables.
2) *Work changes and scope creep*: Ambiguities can lead to additional work from changes, corrections, or rework to meet the owner's expectations. Disputes can arise over the responsibility and liability for associated costs in time and money.
3) *Work quality*: Ambiguities can affect quality if what was interpreted and built is not what is expected by the other party. Quality issues are ready sources of disputes between the parties.
4) *Progress or interim payment disputes*: When there is ambiguity in payment terms and schedules, with undefined milestones and draw intervals, disputes can arise over the timing and amount of payment.
5) *Work and responsibility assignments*: When there is ambiguity in work and responsibility assignments, disputes can arise over consequential delays from relegation of duty and fights over responsibility.

In order to prevent or avoid disputes from ambiguities in contract documents, it is essential to embrace the culture of QA/QC/TQM, Six Sigma, and PDM, and to proactively engage in thorough contract vetting, matching, reviewing, and streamlining relative to the particular project requirements and circumstances. Ambiguities, discrepancies, and inconsistencies in the documents must be identified, clearly defined, and completely cleared before project commencement. Involving a contract attorney and other project stakeholders may help to remove any embodied ambiguities. If any ambiguity, discrepancy, or inconsistency still seeks to show and portend disputes, steps must be taken to swiftly douse them before they escalate to resource-consuming negotiation, mediation, arbitration, or litigation.

Please refer to Table 14.1 for the recommended PDM's D-PAD concept applicable in this regard.

6.3 Discordant Understanding and Interpretation of Contract Clauses

Where contracts have not been written with clear, precise, and definitive words and clauses, they are open to multiple interpretations, which may be discordant among the project participants and stakeholders. This discordance translates into how the project is done, and what constitutes a breach of contract can be continually debatable. Here are some of the ways discordant understanding and interpretation of contract clauses can arise in the project process:

1) *Ambiguity*: Some words used to construct the contract clauses are by nature ambiguous; If not clearly and precisely defined, they are open to multiple interpretations by different parties, and the outcome is likely discordance. Already used for illustration are words or phrases like "reasonable," "adequate," or "or equal."
2) *Disparity of interests*: The disparity in interest on the project between the parties makes it possible for each party to interpret the contract in a way that most aligns with or suits their interest—profits maximization for the contractor and cost-saving for the owner. Such disparity of interests brings about discordant understanding.
3) *Work scope*: The contractor is likely to have the most austere interpretation of the scope of work outlined in the contract, while the owner takes the opposite view. Such discordance can complicate change orders, additional work and costs, as well as rework.
4) *Contrasting view of quality*: The contractor is likely to have a less stringent interpretation of the work quality outlined in the contract, while the owner takes the opposite view. Once more, such discordance can complicate defects and work corrections, as well as rework.
5) *Risk allocation*: Contract clauses that allocate risks to the parties are open to different interpretations by each party in the most favorable application to itself. Contract clauses to do with bonds and insurance requirements, indemnification, and limitations of liability are likely to be interpreted by each party in terms and understanding that are most favorable to it.

Discordant understanding and interpretation of contract clauses are a major problem in the project process that can lead to disputes between the parties in the following instances:

1) *Misaligned expectations*: As the parties want to interpret the contract clauses in terms and understanding that are most favorable to them, this leads to misalignment of expectations that can lead to conflicts and disputes. They can disagree on the deliverables, responsibilities, and liabilities.

2) *Disagreement on work scope*: As alluded to above, while the contractor may have a diminutive view of the work scope, the owner is prone to expanded and often unrealistic apprehension of the scope. This scenario is a fertile ground for disputes between the parties.
3) *Disparate view of quality standards*: While the contractor is likely to have a less stringent interpretation of the work quality outlined in the contract, the owner takes the opposite view. Such discordant views complicate defects and work corrections, as well as rework, and all can lead to disputes between the parties.
4) *Risk allocation disagreement*: Allocation of risks to parties by contract clauses can bring about adversarial and disputatious relationships, especially where such clauses are not clear and definitive, and they concern bonds and insurance requirements, obtaining permits and approvals, indemnification, and limitations of liabilities.

In order to prevent or avoid disputes from discordant understanding and interpretation of contract clauses, it is essential to embrace the culture of QA/QC/TQM, Six Sigma, and PDM, and proactively engage in thorough contract clauses' vetting, reviewing, and streamlining relative to the particular project requirements and circumstances. The clauses should be clearly defined and completely clarified before project commencement. Involving a contract attorney and other project stakeholders may help to shore up the clarity and commonality of the clauses' understanding and interpretation. If any discordant understanding and interpretation of the clauses still seeks to show and portend disputes, steps must be taken to swiftly douse them before they escalate to resource-consuming negotiation, mediation, arbitration, or litigation.

Please refer to Table 14.1 of for the recommended PDM's D-PAD concept applicable in this regard.

6.4 Omissions and Commissions

Omissions and commissions in the context of a contract respectively refer to missing clauses that should have been included and superfluous inclusion of clauses that should not have been included, all of which engender breaches, disputes, and liabilities. In the implementation of the contract, they can also respectively refer to the failure to do what a party to the contract is required to accomplish under it or wrongly doing what a party should not do according to the contract.

Essentially, contracts devoid of omissions and commissions facilitate effective and unambiguous communication, coordination, and elimination or minimization of errors, as well as consequent delays and costs in the project delivery process. Contract clauses are among the primary means of fully conveying the project design and execution intents, and with other accompaniments like drawings, specifications, schedules, and notes, constitute complete and comprehensive information for the successful delivery of the project on time, within budget, and to quality. Negligent contracts replete with omissions or commissions are potent grounds for confusion, conflicts, and disputes due to their impact on time, cost, and quality that are the express and implied contract terms. Omissions mean that vital aspects are oversighted or not considered or included; commissions mean including what should not be included that may ultimately mislead and bring about confusion and deviations from the project design intent.

If the contract is replete with omissions and commissions, it can bring about conflicts and disputes as follows. From the perspective of omissions, conflicts and disputes can be triggered through the following avenues:

1) *Inadequate contract clauses*: Contract clauses may be insufficient in number or extent to cover all grounds required for the accomplishment of the applicable project. For instance, in certain suits of construction contracts, all unforeseen circumstances of the site may not have been fully covered, and a problem arises when such unforeseen situations show up while the project is being executed. This inadequacy can trigger conflicts and disputes.
2) *Contract clause(s) deficiency*: It is possible for some contract clauses to be deficient, meaning that they may not have been in-depth in covering all the possibilities of the clauses. For instance, a contract clause on force majeure may have covered weather and physical conditions but failed to cover situations like a pandemic, as witnessed globally across construction projects and contracts during COVID-19. Such a deficiency is a ready trigger of conflicts and disputes. COVID-19 took most or all affected projects by surprise, leading to contested disputes, which were either resolved amicably or disastrously.
3) *Misinterpretations*: Contracts replete with omissions are prone to misinterpretations or varying interpretations, additions, and subtractions, the correction of which can consume the resources of time, materials, labor, equipment, and money. Conflicts and disputes can arise as a result.
4) *Discrepancies*: The omissions may be discrepant and incongruent to the drawings and specifications, and these are boiling points of incorrect construction that would be costly to correct or rework in terms of the resources of time, materials, labor, equipment, and money. Coordination is difficult under this scenario, and conflicts and disputes can ultimately arise as a result.
5) *Requests for information (RFIs) and change orders*: Contracts with omissions present confusion, challenges, and calls for more RFIs and change orders, all of which can introduce additional complexity and uncertainty as the project progresses, thus altering the scope as well as the intents and objectives of the project in terms of time, cost, and quality. This can bring about conflicts and disputes between the parties.

6) *Delays from regulatory authorities*: Contracts with omissions that fail to consider and accommodate regulatory requirements, codes, and standards attract negative regulatory oversights that can cause delays or ultimately halt the project altogether. This can cause disputes between the parties if the project timeline is affected, with consequential damages and the fight over who is to take the liability and responsibility.
7) *Breach of contract*: As part of the contract documents, contract omissions can facilitate a breach of contract in many ways. Failing to accomplish any aspect of the project as a result of contract omissions constitutes a breach of contract that is a ready trigger of disputes between the parties.

Equally, from the perspective of commissions, conflicts and disputes can be triggered through the following avenues:

1) *Irrelevant contract clauses:* Included contract clauses may be irrelevant for the accomplishment of the applicable project, and their inclusion introduces unwarranted judgment calls and filtering that can bring about confusion and inconsistency. For instance, while using a boilerplate of a suit of contract for different project types, some clauses may apply to some projects but not all projects. Such a scenario presents the challenge of which clauses to consider, which to delete, and who is the party to make that call. This situation can trigger conflicts and disputes.
2) *Unwieldy, ambiguous, dichotomous, or discrepant contract clauses*: The commissions may be unwieldy, ambiguous, dichotomous, or discrepant and incongruent to the drawings and specifications, and these are boiling points of incorrect construction that would be costly to correct or rework in terms of the resources of time, materials, labor, equipment, and money. Coordination is difficult under this scenario, and conflicts and disputes can ultimately arise as a result.
3) *Misinterpretations*: Contracts replete with commissions are prone to misinterpretations or varying interpretations, preferential judgment calls, additions, and subtractions, correction of which can consume the resources of time, materials, labor, equipment, and money. Conflicts and disputes can arise as a result.
4) *RFI and change orders*: Contracts with commissions present confusion, challenges, and call for more RFIs and change orders, all of which can introduce additional complexity and uncertainty as the project progresses, thus altering the scope as well as the intents and objectives of the project in terms of time, cost, and quality. This can bring about conflicts and disputes between the parties.
5) *Delays from regulatory authorities*: Contracts with commissions that fail to consider and accommodate the applicable regulatory requirements, codes,

and standards attract negative regulatory oversights that can cause delays or ultimately halt the project altogether. This can cause disputes between the parties if the project timeline is affected, with consequential damages and the fight over who to take the liability and responsibility.

6) *Breach of contract*: As part of the contract documents, contract commissions can facilitate a breach of contract in many ways. Wrongly accomplishing any aspect of the project as a result of contract commissions constitutes a breach of contract that is a ready trigger of disputes between the parties.

In order to prevent or avoid disputes from contracts replete with omissions and commissions, they should not be used to break ground on the project to begin with. They must be thoroughly reviewed, edited, and made to be suitable and sufficient to successfully deliver the project's design intents and objectives on time, within the budget, and to quality. QA/QC/TQM, Six Sigma, and/or ultimately PDM principles and precepts should be embraced to ensure that the contracts are complete and comprehensive to communicate precisely and effectively to lead the way to a successful project outcome. Where any contract omission and commission issues arise nonetheless, swift action should be taken to respond to them to avoid any escalation to conflicts and disputes.

Please refer to Table 14.1 for the recommended PDM's D-PAD concept applicable in this regard.

6.5 Sharing of Risks and Rewards

Construction projects come with a lot of risks and rewards. The risks and rewards are shared between the parties through a contract arrangement. How they are shared and implemented can determine how adversarial the parties can be throughout the project process. Here are the modalities of risks and rewards shared in the project process:

1) *Owner share*: While the owner is rewarded by the successful and hitch-free completion of the project, it deals with the risks of project financing, regulatory, and environmental requirements, as well as the overall project performance risks relative to its purpose: for instance, fitness for purpose, and revenue generation.
2) *Contractor share*: While the contractor is rewarded with compensation and profits from the project as well as reputation-building, it deals with the risks of time, cost, quality, people, and safety management in the process of delivery.
3) *Subcontractors and suppliers' share*: Very similar to that of the contractor relative to their work aspects and the human, material, and equipment resources to deliver them.

4) *Design professionals' share*: Design professionals are rewarded by compensation for their design services, and they deal with the risks of quality, completeness, and comprehensive documentations, errors, omissions, commissions, and general negligence.

Appropriately sharing risks and rewards through contractual arrangements is crucial in ridding construction project delivery of adversarial relationships between the parties. If project risks and rewards are not properly allocated, they can bring about disputes due to the following reasons:

1) *Inappropriate risk and reward sharing*: When risks and rewards are inappropriately shared or distributed, disputes can arise between the parties. For instance, if the risk of unforeseen site conditions is allocated to the contractor on top of its compensation, and those site's conditions turn out to be far worse than their perceptions in the agreed-upon contract. Disputes can arise if the unforeseen site conditions cost more and eat into the contractor's compensation or reward.
2) *Inherently adversarial contracts*: Except for the claims of the newer integrated project delivery (IPD) contracts that tout collaborative shared risks and rewards between project owners and contractors, most traditional contracts are said to be adversarial, with each party separately defending its positions rather than collaboratively. This adversarial nature makes them inherently susceptible to triggering disputes between the parties.
3) *Work scope creep and changes*: Work scope creep and changes can alter the balance and distribution of shared risks and rewards due to added responsibilities and risk exposure, and this can lead to disputes between the parties.
4) *Risk shifting without commensurate compensation*: When one party fails in its responsibility for an allocated risk, risk may shift to the other party without necessarily commensurately compensating for it. For instance, if the contractor fails to pay the subcontractor or the supplier, the owner is put at risk of a lien on the project by the unpaid subcontractor or supplier. This can lead to disputes between the parties.

In order to prevent or avoid disputes from contractual risks and rewards sharing in construction projects, it is essential to be circumspect over the form of contract adopted. Perhaps such disputes can be prevented by using the more modern IPD contracts that have been touted as nonadversarial, but embodiments of collaborative shared risks and rewards between the parties would ensure that there is a balance and fair distribution. The party that should better bear a certain risk than the other should accept to take charge of that. Helpful risk assessment and management best practices should be embraced, as well as the inclusion of risk management methods and requirements such as bonds, insurance, indemnifications, limitations to liability clauses, disclaimers, and contingency allowances.

By clearly and proactively defining and delineating each party's rewards, responsibilities, and liabilities, the potential of disputes is mitigated considerably, if not eliminated completely. Steps must be taken to swiftly douse any portending disputes before they escalate to resource-consuming negotiation, mediation, arbitration, or litigation.

Please refer to Table 14.1 for the recommended PDM's D-PAD concept applicable in this regard.

Chapter Review Questions

1) Ambiguities in contract documents can lead to disputes due to multiple interpretations. True/False?
2) All construction contracts are inherently collaborative and prevent disputes. True/False?
3) Omissions in a contract refer to clauses that should have been included but were not. True/False?
4) Commissions in a contract refer to the failure to fulfill contractual obligations. True/False?
5) Sharing of risks and rewards in contracts can influence the likelihood of disputes. True/False?
6) Poorly defined contract clauses can result in misinterpretation and disputes. True/False?
7) Contracts with clearly defined roles and responsibilities help prevent disputes. True/False?
8) Disputes over risk allocation are uncommon in construction contracts. True/False?
9) The inclusion of clauses that contradict other parts of the contract creates ________ in the agreement.
10) Contracts that do not reflect the specific needs of a project may be described as ________.
11) The legal agreement between two parties in a construction project is called a ________.
12) The process of identifying and mitigating potential dispute triggers before project commencement is known as ________.
13) When contract clauses are interpreted differently by each party to suit their own interests, it can lead to ________ in understanding and implementation.
14) A contract clause that is vague or open to multiple interpretations is considered ________.
15) The failure to allocate project risks and rewards equitably between parties can result in ________ relationships and potential disputes.

16) Which of the following is a common cause of contract-related disputes?
    a) Clear communication
    b) Explicit contract clauses
    c) Ambiguities in the contract
    d) Defined responsibilities
17) What is a recommended strategy to prevent disputes from contract ambiguities?
    a) Avoid reading the contract
    b) Use verbal agreements
    c) Work with unambiguous and explicit contracts
    d) Ignore discrepancies
18) Which of the following is NOT a benefit of working with clear contracts?
    a) Improved relationships
    b) Increased efficiency
    c) Higher risk of disputes
    d) Greater enforceability
19) Which of the following is a key aspect of a construction contract?
    a) Project location
    b) Project timeline
    c) Weather forecast
    d) Site photos
20) Which contract type is considered more collaborative in risk and reward sharing?
    a) Lump-sum contract
    b) Cost-plus contract
    c) Integrated project delivery (IPD)
    d) Unit-price contract

Exercises

Exercise 6.1 Contract Clause Clarity Audit

Objective: Identify and evaluate ambiguous clauses in a real or sample construction contract.

Instructions:

- Obtain a sample construction contract (or use a real one with permission).
- Review the contract and highlight any clauses that may be ambiguous, vague, or open to multiple interpretations.

- For each clause identified:
  - Explain why it may lead to disputes.
  - Suggest a revised version of the clause that improves clarity and reduces ambiguity.
- Present your findings in a short report or group presentation.

Outcome: Develop skills in identifying and resolving potential dispute triggers in contract language.

Exercise 6.2 Risk and Reward Allocation Simulation

Objective: Analyze how risk and reward sharing in contracts affects project relationships.

Instructions:
- Form small groups and assign roles (owner, contractor, subcontractor).
- Each group receives a hypothetical project scenario with a contract that includes risk and reward clauses.
- Review the contract and discuss:
  - Which risks are allocated to each party?
  - Are the rewards proportionate to the risks?
  - What potential disputes could arise from the current allocation?
- Propose a revised risk/reward structure that promotes fairness and minimizes conflict.

Outcome: Understand the impact of contract structure on project dynamics and dispute potential.

Discussion

1) How does the precision or vagueness of contract language influence the likelihood of disputes in construction projects? Share examples of how ambiguous clauses have led to misunderstandings or legal conflicts.
2) Traditional construction contracts are often described as adversarial. Do you agree with this characterization? How might integrated project delivery (IPD) or other collaborative models change the dispute landscape in construction?

7

Attitude and Disposition-Related Dispute Flashpoints

Readers' Learning Objectives and Outcomes

1) *Analyze the various flashpoints and triggers of conflicts and disputes in the industry related to the attitude and disposition of the parties, participants, and stakeholders.*
2) *Conceptualize plausible prevention, avoidance, or dousing approaches that may be applicable to these conflicts' and disputes' flashpoints and triggers.*
3) *Match and apply prevention, avoidance, or dousing strategies effective or suitable to each conflict's and dispute's flashpoints and triggers.*

7.1 Introduction

Considering the many participants and stakeholders in the construction process, it can be rightly expected that attitudes and dispositions would differ multifariously. Attitudes and dispositions refer to the behavioral and personal traits of the participants and stakeholders. The differences in attitudes and dispositions center around the behaviors, expectations, and perspectives of the participants and stakeholders, and here are some of the fueling factors:

1) *Communication difficulties*: Differing behaviors, expectations, and perspectives would engender communication difficulties among project participants and stakeholders. Steps should be taken to align these aspects of differences, militating against clarity and effectiveness of communication.
2) *Differing interests, objectives, and priorities*: The participants and stakeholders may have differing interests, objectives, and priorities. For instance, the project owner may prioritize quality at the lowest cost, while the contractor prioritizes maximization of profit even at the sacrifice of quality.
3) *Clash of personalities*: The many project participants and stakeholders come with different, and often conflicting, personalities that would affect their working style and relationship with one another. These conflicting characteristics can affect communication, team dynamics, and overall project performance.
4) *Trust issues*: Project participants and stakeholders with clashing personalities would find it difficult to trust and partner with one another for the success of the project. With a lack of trust, project participants and stakeholders readily question one another's motives, decisions, and actions.
5) *Cultural and organizational differences*: There may be cultural differences among project participants and stakeholders, and there can be differences from one participating organization to the other. Such differences can show up in work norms, communication practices, decision-making, and implementation. These differences can only bring friction in the process.

Attitudes and dispositions of project participants and stakeholders are important to consider, as they dictate some conflicts and disputes that can arise in the construction delivery process and how they escalate. The dictate participants and stakeholders' assertiveness and aggressiveness or lack thereof, communication style, trust and distrust, decision-making and implementation approach, flexibility and rigidity, time management, stress and pressure handling, problem-solving style, and regard to expertise and roles. As a result, the participants' and stakeholders' litigious nature, cultural disposition to work and contracts, communication, and team-playing styles would differ as much as they come. These differences are further discussed hereunder.

7.2 Litigious Nature of Parties and Participants

The litigious nature of project participants and stakeholders (owners, architects, designers, engineers, consultants, contractors, subcontractors, suppliers, sub-subcontractors, etc.) refers to their tendency to resort to disputes and legal actions of litigation or quasi-judicial systems, such as mediation and arbitration. Some are more litigious than others. The litigious nature of project participants and stakeholders is controlled by their differing attitudes and dispositions that have been discussed above, and may be fueled by and predicated on the following:

1) *Complexity and risk in the project*: Construction projects are inherently complex and risky, and the more complex and riskier the project, the higher the potential for litigable conflicts and disputes. High stakes are in financial investments, timelines, and quality expectations, and this heightens the sense of risk and means to protect the interests of the parties.
2) *Number of participants and stakeholders*: The more the number of project participants and stakeholders, the more the likelihood of litigable conflicts and disputes.
3) *Location*: Some project locations may facilitate litigable conflicts and disputes more than others. For instance, litigable conflicts and disputes may be encouraged or discouraged depending on the state of the project location and whether it is a right-to-work state or not. The application of the rule or law can vary from state to state and country to country.
4) *Contract motivation*: Some contract forms are more adversarial than others, and they facilitate rather than dissuade litigable conflicts and disputes. For instance, lump-sum contracts are said to be more adversarial than integrated project delivery (IDP) arrangements, which are seen as collaborative with equitably shared risks and rewards.
5) *Absence of conflicts- and disputes-minimizing and dissuading factors*: The litigious nature of project participants and stakeholders is burnished by the absence of conflicts and disputes, minimizing and dissuading factors such as QA/QC/TQM, Six Sigma, alliancing and partnering, as well as PDM.

The litigious nature of project participants and stakeholders can trigger conflicts and disputes in the following circumstances and ways:

1) *Contractual disputes*: There are many forms of contracts used in the project delivery process. Some can encourage noncollaborative relationships, and can be described as adversarial, with each party seeking to protect its interests. The contracts can be interpreted differently, and the litigious nature of project participants and stakeholders, rather than swiftly deescalating minor issues in no time, motivates resorting to time- and resources-consuming litigation or quasi-judicial systems, such as mediation and arbitration.

2) *Work change issues*: The litigious nature of project participants and stakeholders drives them to escalate even the simplest work change issues to litigation or quasi-judicial systems, such as mediation and arbitration.
3) *Time extension arguments or disagreements*: The litigious nature of project participants and stakeholders drives them to escalate time extension arguments or disagreements to litigation or quasi-judicial systems, such as mediation and arbitration.
4) *Payment issues*: The litigious nature of project participants and stakeholders drives them to escalate even the minutest payment issues to litigation or quasi-judicial systems, such as mediation and arbitration.
5) *Quality and performance issues*: The litigious nature of project participants and stakeholders drives them to escalate quality and performance issues to litigation or quasi-judicial systems, such as mediation and arbitration.
6) *Liability and negligence claims*: Such claims will be rife in the process, depending on the litigious nature of project participants and stakeholders; those can only lead to litigable conflicts and disputes.

In order to prevent or avoid the litigious nature of project participants and stakeholders from readily bringing about disputes in the process between the parties, there must be a proactive recognition and provision or accommodation of this reality through various means, such as contracts, management practices, open and clear communication channels, transparency, and fostering a collaborative and mutual respect for culture. Where conflicts and disputes portend to insist to arise, swift steps must be taken to deescalate them at no cost of time and resources.

Please refer to Table 14.1 for the recommended PDM's D-PAD concept applicable in this regard.

7.3 Cultural Disposition to Work

Cultural disposition to work refers to collective morals, norms, values, attitudes, and behaviors of project participants and stakeholders, which can be multifarious and shape the work environment and practices that influence how projects are planned, managed, and executed. Several factors characterize cultural disposition to work in construction, and they include:

1) *Work ethic and capacity*: Work ethic and commitment to overcome all challenges and get the job done efficiently and properly are dictated by cultural disposition to work, which can vary among project participants and stakeholders.
2) *Working collaboratively*: Cultural disposition to work can also determine whether project participants and stakeholders work collaboratively to achieve

project objectives with few or no issues. Some dispositions encourage or discourage teamwork.

3) *Management of risk and safety*: Cultural disposition to work can also determine how project participants and stakeholders approach and handle risk and safety issues. Some dispositions are carefree or cautious about risk and safety issues.
4) *Adaptability and flexibility*: Cultural disposition to work can dictate how adaptable and flexible project participants and stakeholders are to changes and challenges, such as work changes, adverse weather, and supply chain disruptions. For instance, some project participants and stakeholders may be a little disrupted in their work by snow days and other nonextreme weather conditions.
5) *Commitment to craftsmanship and quality*: Commitment to craftsmanship and quality can be determined by the cultural disposition of project participants and stakeholders to work. Some project participants and stakeholders can deliver fairly consistent craftsmanship and quality stemming from their cultural disposition to work, which may vary from one region to another or from one country to another.
6) *Integrity and professionalism*: These attributes can vary relative to the differing cultural dispositions of project participants and stakeholders to work. This will dictate their adherence to ethics, standards, regulations, and contractual obligations.

Recognizing, adapting, and adjusting to varying cultural dispositions to work are essential to successful project outcomes through appropriate participant and stakeholder engagement, management, and fostering of a positive, collaborative work atmosphere. The cultural disposition of project participants and stakeholders to work, when not properly leveraged and managed in the process, can bring about disputes in the following circumstances:

1) *Clash of work ethics*: The cultural disposition of project participants and stakeholders to work can present a clash of work ethics. In extreme situations, project participants and stakeholders may have differing expectations and views of work hours and days, pace and urgency of work, and commitment to meet deadlines. For instance, some may take off on Friday or commit work hours to religious observances, which can be at variance with the collective project goal of timeliness and cost-effectiveness. When some are not pulling their weight to meet the project objective of timeliness, disputes can arise in this scenario.
2) *Communication and collaboration*: Cultural disparity can dictate how project participants and stakeholders communicate, and how open they are to collaborative work and problem-solving. This can influence their clarity and directness in succinct communication. Hierarchy and egalitarianism may

exist and determine how they communicate and make decisions, or whether communication and interactions will be formal or informal, all of which can affect the effectiveness and comprehension of communication. Where ineffective communication or miscommunication leads to costly mistakes or liabilities, disputes can arise.

3) *Differential handling issues*: Cultural norms can dictate how issues are handled or managed in the project process. While some would handle the issues swiftly and decisively, others may prioritize subtle approaches to maintain harmony and avoid confrontations. Where issues that need to be handled swiftly and decisively are not so approached, and costly mistakes or liabilities are incurred, disputes can arise.
4) *Respect for authority, role, and expertise*: Issues related to respect for authority, roles, and expertise in the decision-making, implementation, and general piloting of the project can bring about disputes triggered by differing cultural dispositions to work, tradition, and expertise.
5) *Multiculturalism*: Construction projects are often comprised of participants and stakeholders of multicultural backgrounds and perspectives. While multiculturalism can bring invaluable assets to the project, if not properly harnessed and leveraged, it can also be a trigger for disputes due to the clashing backgrounds, perspectives, and approaches. Language barriers, cultural biases, and misunderstandings may be at play and raise tensions that are harbingers of conflicts and disputes.
6) *Trust and interpersonal relationships*: Cultural disparity and disposition to work can affect the level of trust and interpersonal relationships that can influence successful project outcomes. Trust is crucial in collaboration, decision-making, and implementation. Disputes can arise over real or perceived breaches of trust, conflicting cultural perspectives, and bridging cultural divides on a project.

In order to prevent or avoid disputes from differing cultural dispositions to work, there must be concerted proactive efforts at ascertaining, recognizing, and providing for or accommodating the disparities through various means, such as cultural awareness and training, culture-sensitive communication approaches, adoption of diversity and inclusion principles, contract and management measures, transparency, and fostering of a collaborative and mutual respect for culture and harmonization. Where conflicts and disputes portend to insist to arise, swift steps must be taken to deescalate them at no cost of time and resources.

Please refer to Table 14.1 for the recommended PDM's D-PAD concept applicable in this regard.

7.4 Cultural Disposition to Contracts

Cultural disposition to contracts refers to collective morals, norms, values, attitudes, and behaviors of project participants and stakeholders, which can be multifarious and shape how they perceive, negotiate, and execute contracts on projects. Several factors characterize cultural disposition to contracts in construction, and they include:

1) *Adherence to contractual obligations*: Cultural disparities can dictate the extent to which project participants and stakeholders respect agreements or contracts, whether written or oral. Fulfilling contractual obligations in an agreed-upon, timely, and satisfactory manner may all subsume to the differing cultural dispositions of the project participants and stakeholders. Some may strictly emphasize written agreements as the only legally binding undertaking outlining the terms, conditions, obligations, and responsibilities of the parties. In other words, an agreement that is not in writing is not legally binding. Others may mutually accept any agreement, oral or written, as legally binding. These cultural disparities must be proactively uncovered and adjusted to in the project delivery interactions and relations.
2) *Risk and liability allocation*: Cultural disparities may determine how project participants and stakeholders process the allocation of risks and liabilities, as well as the sharing of rewards. In other words, they dictate their openness to collaborative problem-solving and partnering. While some cultures may be rigid about strict risk and liability allocations, others may be open to some flexibility, with all hands on deck to achieve the collective project goal.
3) *Trust and relationship building*: Cultural disparity in the views of contracts can dictate the level of trust and relationships among project participants and stakeholders that can influence successful project outcomes. Trust is crucial in collaboration, which is central in relationship building, collective decision-making, and implementation of projects. Any contract that fosters this ambiance facilitates successful project outcomes, but cultural disparities in the project participants' and stakeholders' views of it are cogs in the wheel here.
4) *Negotiation*: Cultural disparity in the views of contracts can influence project participants' and stakeholders' approach to negotiation and bargaining matters, which may sometimes become imperative in the process. The need to concede, compromise, or seek consensus and achieve a balance of interests among the parties may be hindered by the differing cultural disposition to contracts by the project participants and stakeholders.

Uncovering and timely understanding, adapting, and accommodating differing cultural dispositions of project participants and stakeholders are essential to successful project outcomes. Not doing so can contribute to disputes under certain circumstances that include:

1) *Contract understanding, interpretation, and implementation*: Cultural differences may influence how parties understand, interpret, and implement contract terms, obligations, and responsibilities. Misunderstandings or misinterpretations stemming from lingo and language disparities, nuances, norms, and legal interpretations can trigger disputes over the intent of contractual provisions on the extent and scope of work. For instance, what constitutes legitimate and excusable causes of delay may be differently understood or interpreted by the parties, and disputes can arise over responsibility for damages and liabilities for delays.
2) *Negotiated contract issues*: Some cultures may have more tendency toward hard contract bargaining and a win-at-all-cost approach. It means there is no balance of the interests of the parties, and one party may feel unfairly pressured and disadvantaged from the start. Where such a stance doesn't jell with the other party but is perceived as too aggressive, adversarial, and uncompromising, the seed of distrust and friction is sown that will be a flashpoint or lightning rod for disputes.
3) *Inequitable allocation of risks, liabilities, and rewards*: Some cultures may not emphasize balance and equitability in the contractual allocation of project risks, liabilities, and rewards, where a win-at-all-cost approach by either party is the norm. Disputes can arise if one party feels unfairly overburdened by emergent risks and liabilities while being denied commensurate rewards.
4) *Enforcement and compliance with contractual obligations*: Some cultures may emphasize strict enforcement and compliance with contractual obligations and assertion of rights, and they are quick to dispute even the slightest noncompliance by another party. In contrast, other cultures may be more flexible to concede, compromise, or informally resolve issues rather than escalate them and affect the progress of the project. Such differing expectations and views of enforcement and compliance can be tension and dispute flashpoints or triggers between the parties.
5) *Tense or relaxed relationship dynamics*: While contracts formalize the legal relationships and obligations of the parties, they also shape interpersonal relationships among or between them. Cultural differences can dictate how much tense or relaxed relationship dynamics exist among the parties as they navigate contractual and project issues. Of course, tense relationship dynamics are potential triggers of disputes between the parties.
6) *Challenged effective communication*: When effective communication is challenged by cultural differences stemming from language, lingo, norms, and

values, disputes can arise where there is resultant poor communication and understanding of contractual obligations, responsibilities, and performance, as well as issues of project changes, documentations, and reporting.

7) *Diverse perspectives on handling issues*: Some cultures may emphasize a formal approach to handling project issues, which suggests that the parties may resort to formal negotiations, arbitration, or negotiation as strictly provided in the contract. The scenario triggers disputes between the parties on even the minutest issues if they fail to align on a collectively preferred approach to handle the issues outside of contestations.

In order to prevent or avoid disputes from differing cultural dispositions to contracts, there must be concerted proactive efforts at uncovering, recognizing, understanding, and providing for or accommodating the disparities through various means, such as cultural dispositions awareness and analysis, culture-sensitive communication approaches, adoption of diversity and inclusion principles, contract and management measures, transparency, and fostering of a collaborative and mutual respect for culture and harmonization. Where conflicts and disputes portend to insist to arise, swift steps must be taken to deescalate them at no cost of time and resources.

Please refer to Table 14.1 for the recommended PDM's D-PAD concept applicable in this regard.

7.5 Communication and Miscommunication

Communication refers to the act of exchanging information between or among parties. This exchange of information may be clear and succinct but if not so due to any reason, miscommunication is said to take place. Hence, information can be both clearly and succinctly exchanged or unclearly and clumsily exchanged. On a project, either has critical implications for the successful or unsuccessful outcomes, especially due to the many participants and stakeholders involved. There can sometimes be disparities in transnational communications on remote and overseas projects. For instance, a "Yes" in Western communication may mean a definite consent, but for someone with an unadulterated Japanese cultural perspective, a "Yes" answer may not mean absolute consent but rather a motion or a signal to continue what you are conveying or saying in communication. To some, a "Yes" is a means to hurry and end the conversation rather than an acceptance of what is conveyed to be implemented.

Essentially, on any project, several means of communication apply, and they include:

1) *Graphics*: This constitutes the use of drawing illustrations to communicate the project concept and intent. That is, drawings are used to communicate

those aspects of the project that are better conveyed through graphical media. The graphics can be hard copies or electronic and constitute drawing documents or construction documents as applicable to building projects.

2) *Dimensions, notes, and schedules*: As graphics alone can be inadequate to completely convey project concept and intent, they are supplemented by dimensions, notes, and schedules.
3) *Specifications*: These refer to the communication of project concepts and intent in written rather than drawn form, whether supplemented by dimensions, notes, and schedules. That is, words are used to communicate those aspects of the project that are better conveyed through written media.
4) *Verbal lingo and language*: Communication can be through professional lingo and may be in different languages. However, the construction site is often peopled by workers of different backgrounds and cultures, and language barriers, different nuances, and understandings can affect communication and how clearly it is transmitted, understood, and implemented, just as in the Japanese example above.
5) *Miscellaneous communication*: Other communication means include shop drawings, bulletin drawings, change and site directives, change orders, site orders, letters, reports, minutes, diary logs, permit slips, safety postings, and hazard communication with safety data sheets (SDS), payment certificates, and invoices. Any imprecision in any of these can lead to miscommunication.

Project concept and intent can be completely and accurately conveyed through the collective of the above communication means. However, that's sometimes not the case, and miscommunication can occur and bring problems due to the following reasons:

1) *Negligent errors, omissions, and commissions*: The drawing documents may have been sketchy at best, erroneous, missing vital graphic communication information and details, and failing to fully convey the necessary graphical aspects of the project. This problem can extend to dimensions, annotations, and schedules, as well as specifications. Negligence may stem from the experience and skill of the designer.
2) *Discrepancy*: Discrepancy happens when the drawing documents and specifications say slightly or completely different things and can be interpreted differently.
3) *Ambiguity*: Communication, graphical or written, can be ambiguous with its accompanying problems. It could be conveying what can be interpreted and implemented differently by different people, which is why clarity and precision are crucial.

Clear and succinct communication is essential in ensuring project success. If not clear and succinct, miscommunication can occur and bring about disputatious errors, omissions, changes, rework, loss of productivity, jeopardized safety, and claims. When communication is ineffective, and miscommunication is the result, the following dispute issues can arise:

1) *Dispute from incomplete and/or inaccurate communication*: If a project owner communicates a change verbally and perhaps incompletely, and this is not properly and accurately documented or confirmed, the contractor may proceed based on a different understanding or assumption, and this can lead to disputes over responsibility and cost of the wrong implementations.
2) *Dispute from ambiguous communication*: When projects are not detailed adequately and/or the specifications are ambiguous, they will be open to interpretations by different parties who may proceed with wrong assumptions to deliver work that is below expectations, which ultimately results in disputes.
3) *Dispute from information correction time and delays*: Time will be expended to correct or make up for incorrect information communication, and disputes can arise over which party takes responsibility for any associated costs of consequential delays, disruptions, and damages.
4) *Dispute from communication documentation*: When there is miscommunication due to inadequate communication documentation, disputes may arise from conflicting recollections over what was agreed upon or done.
5) *Dispute from coordination difficulties*: Miscommunication can hamper the coordination of the project participants and stakeholders, and this can negatively impact the project timelines, costs, and quality outcomes that would lead to disputes.
6) *Dispute from lingo and language differences*: Where miscommunication happens due to lingo and language differences, as well as culture, norms, and nuances among project participants and stakeholders, disputes can arise over misunderstandings of project execution and expectations, compliance to standards and regulations, as well as contractual obligations, all of which can lead to disputes from the impacts on project timelines, costs, and quality outcomes.
7) *Dispute from relationships*: Where communication is ineffective, and miscommunication is rife, relationships among project participants and stakeholders will be strained, and such a scenario is a potential trigger for conflicts and disputes from matters that should not be issues.

In order to prevent or avoid disputes from communication or miscommunication, there must be a culture of, and a concerted proactive effort at, thoroughly analyzing and scrutinizing the communication instruments, channels, and protocols of the project from start to finish. Documentation must be complete and up to date. Culture-conscious communication must be adopted in recognition of the often multicultural makeup of the project participants and stakeholders. Hence, adoption of diversity and inclusion principles, contract and management measures, transparency, and fostering of a collaborative and mutual respect for culture and harmonization are all essential. Project participants and stakeholders should be made to feel comfortable asking questions and seeking clarifications. Training in effective communication skills and cultural communication sensitivity should be integrated into project participants' and stakeholders' organizations, and up-to-date technology should be deployed to facilitate real-time communication and collaboration. Where conflicts and disputes portend to insist to arise, swift steps must be taken to deescalate them at no cost of time and resources.

Please refer to Table 14.1 for the recommended PDM's D-PAD concept applicable in this regard.

7.6 Many Players on the One Project

In construction projects, there is a diverse range of participants and stakeholders involved in planning, designing, building, and managing them from conceptual to finished stages and handover or turnover. These participants and stakeholders contribute their resources, expertise, and responsibilities through the phases of the project cycle from conceptual start to finish. The key project participants and stakeholders include:

1) *Project client or owners*: Project owners initiate them and are responsible for their purpose and financing. They define the requirements and program, approve the concept and design, and pay progress and final payments.
2) *Architects and designers*: Architects and designers conceive and design the building project according to the requirements, functional needs, and aesthetic needs of the owner. They prepare the schematic and working drawings, schedules, and specifications that guide the project's construction, and they may also be involved in the oversight to ensure that what has been proposed is what is built. Landscape architects are specialist architects in the horticultural and surrounding aspects of the building, such as planting of trees, turf, mounds, swales, and landscape furniture and other features, but they can also be involved in interior planting. Interior designers deal with designing interior ambience, furniture, lighting, interior space accents, colors, and decorations.
3) *Engineers*: There are many engineers who can be involved in a project, each of whom provides and communicates the necessary project design information, including:
    a) *Land surveyors*: These are those who map the property boundaries, topography, and site features, including its cardinal situation relative to the North and South Poles, or the four cardinal directions of north, east, south, and west.

b) *Geotechnical and soil engineers*: Geotechnical engineers are sometimes called soil engineers, and they are types of civil engineers who specialize in evaluating the characteristics of the soil or site upon which the project will be built, providing information on the bearing capacity of the soil and the water table at the location.
c) *Civil engineers*: These are those who manage site preparation and infrastructure, such as stormwater drainage and retention structures. They deal mainly with the horizontal construction aspects of the building project, such as driveways, roads, paved walkways, bridges, dams, and parking lots.
d) *Mechanical engineers*: These are those who design and handle heating, ventilating, and air-conditioning (HVAC), as well as plumbing, rainwater discharge, and firefighting systems and installations.
e) *Electrical engineers*: These are those who design and handle the electrical installation of the project, including power and fixtures, internal and external.
f) *Cost engineers*: Called quantity surveyors in some climes, these are those who dissect and cost the project, and they also help in cost control throughout the project construction. They provide both priced and unpriced bills of quantities that are helpful in the bid or tendering process and award.
g) *Communication engineers*: These are those who design and handle communication equipment and systems installations of the project.
h) *Safety engineers*: Sometimes called safety officers or managers, are those who design and deploy the safety procedures and protocols to ensure that the project is safely delivered, with participants and stakeholders protected from injury, illness, and property damage.

4) *General contractors (GC)*: Otherwise referred to as "prime contractors" or simply as "contractors," they oversee the whole construction process from site preparation to actual construction and handover, coordinating the work and contributions of all participants and stakeholders to safely deliver the project on time, within the budget, and to quality. The GC coordinates all construction aspects, including procurement, logistics, and quality control.
5) *Subcontractors*: These are those who specialize in different trades and aspects of the project, such as plumbing, masonry, carpentry, electrical, ironmongery, concreting, waterproofing, and roofing. They assist the GC to accomplish specialized project tasks within their expertise areas.
6) *Suppliers and vendors*: These are those who supply different materials, equipment, and components used in the project. The supplies include lumber, rebars, sand, gravel, concrete, doors, windows, plumbing and electrical fixtures, roofing sheets, and tiles.

7) *Construction or project managers (PM)*: These are those who coordinate the entire construction of the project on behalf of either or both of the owner and the GC. They facilitate communication among project participants and stakeholders, coordinate all trades' activities, monitor progress relative to the project schedule, manage the budget, and ensure that the project is built to the expected standard and quality. PMs may have a hierarchy of assistants working for and under them.
8) *Project superintendents or supervisors*: These are those who are hands-on at the site, overseeing and coordinating the day-to-day work of the trades and the direction of project operations at the site on behalf of the GC. Superintendents may have a hierarchy of assistants working for and under them.
9) *Project/job/field engineer*: This is sort of a trainee project manager or superintendent, a starting position for a graduate of construction management, construction engineering or science, or civil engineering who assists and reports to the PM or superintendent on project scheduling, project measurement and reporting, progress billing, cost studies, keeping job logs or records and reports, observations, testing, job engineering and surveys, safety and first aid, as well as payroll time sheets.
10) *Craft foreman or foreperson in charge*: This is a hands-on skilled worker who oversees the work of a specific trade on behalf of the superintendent or subcontractor.
11) *Owner's project representative or clerk of works in other climes*: Otherwise called resident project representative, resident architect, resident engineer, or contracting officer, they are the eyes and ears of the owners or their architects, designers, or engineers who are only on site intermittently. Duties include:
    - monitoring, inspections, and quality control
    - receiving and checking shop drawings, brochures, and samples
    - keeping job diary or reports
    - recording the weather, which trades are at the site and which visitors came to the site on any day
    - receiving and checking pay requests
    - receiving site directives and processing change orders
    - making location surveys
    - receiving and checking or testing materials and supplies
    - performance testing
    - measure of work quantities
    - providing status reports
    - attending project meetings and taking notes or recording as allowed
    - gathering and preparing as-built drawings information, and
    - other needed errands and functions of the project owners or their architects, designers, or engineers.

12) *Construction workers*: These are those who do the various specialized project tasks hands-on on behalf of the trades. They may include skilled workers, otherwise called journeymen, laborers, helpers, trainees, and apprentices. They perform physical tasks, such as excavation and trenching, framing, plumbing, electrical installations, carpentry, roofing, and finishing.
13) *Specialist consultants and experts*: Depending on the scope, purpose, and aspirations of the project, there may be involved a coterie of specialist consultants and experts, such as code consultants, environmental engineers, and consultants on sustainability and environmental assessments, acoustics consultants, fire protection and safety consultants, expert project estimators, and schedulers. They provide expert guidance to support project decision-making and implementation.
14) *Regulatory authorities and inspectors*: Municipal building departments, zoning officers, environmental agencies, and safety inspectors oversee compliance with standards and building codes, zoning regulations, environmental standards, and Occupational Safety and Health (OSHA) requirements. They issue the necessary permits, inspect and endorse compliance, and ensure that construction activities adhere to and meet the required legal and regulatory standards.
15) *Financial institutions and loan insurers*: Financial institutions like banks and credit unions provide funds, as well as construction and mortgage loans, for the projects. In the United States, mortgage loans are principally provided to financial institutions by private and government lenders that include Fannie Mae (Federal National Mortgage Association), Freddie Mac (Federal Home Loan Mortgage Corporation), and Ginnie Mae (Government National Mortgage Association). Mortgages are insured here by Federal Housing Administration (FHA) and Veteran Administration (VA). Construction loans are novated to mortgage loans on completion of construction.
16) *Surety companies*: Surety companies issue project bonds, such as bid bonds, construction bonds that include performance bonds, and payment bonds or labor and material bonds, warranty or maintenance bonds, fidelity bonds, subcontract bonds, treasury bonds, bonds to release retainage, bonds to discharge liens or claims, bonds to indemnify project owner against liens, bonds to protect owner of rented equipment or leased property, judicial or court bonds, license or permit bonds, termite bonds, subdivision bonds, union wage bonds, and self-insured worker compensation bonds. While private construction contracts use common law bonds that include performance and payment bonds, public projects go with statutory bonds, which are established by law with provisions set forth by statutes.
17) *Insurance companies*: Insurance companies assume project risks by providing financial covers and support for projects. Some of the relevant insurance

policies that are taken out on construction projects include those obligated by law, namely, workers' compensation insurance, builder's risk insurance (all-risk and named-peril), and comprehensive or general liability insurance. Builder's risk insurance is a type of property insurance while the building is under the control of the builder or the contractor against incidents such as water damage, fire, explosion, vandalism, theft, and other building-related risks and losses. Depending on the scope and realities of the project, this property insurance for projects under construction can be broad and go beyond builder's risk to as far as earthquake insurance, bridge insurance, steam boiler and machinery insurance, installation floater insurance such as that used for HVAC, property insurance on contractor's own premise (e.g., premises insurance, equipment or floater insurance, motor truck insurance, transportation floater insurance, crime insurance, employee dishonesty insurance or fidelity insurance, valuable papers destruction insurance). Other policies include liability insurance or defense coverages (e.g., public liability and property damage insurance, protective public and property damage liability insurance, completed operations liability insurance, contractual liability insurance, professional liability or indemnity insurance, employer's liability insurance, owner's protective liability insurance), employee insurance (e.g., employee benefit insurance for fringe benefits, Social Security insurance, unemployment insurance, disability insurance), automobile insurance for both damage or loss of contractor's vehicle and liability for bodily injury to others or damage to property, business/accident/life insurance (e.g., business interruption insurance, sole entrepreneurship/proprietorship insurance, key person/employee insurance, business or corporate continuity insurance). Further insurance policies that can be found during project construction include umbrella excess liability insurance, wrap-up insurance, nonoccupational disability insurance, subcontractors' insurance, and group insurance. Projects must ensure to take out and work with the necessary, verifiable insurance certificates for all participants and stakeholders.

18) *Attorneys and related personnel such as mediators and arbitrators*: It is a wise measure to retain attorneys to review and advise on contracts and even specifications. Mediators and arbitrators are on hand to help mediate and arbitrate disputes where they escalate, but PDM seeks to disrupt this norm by proposing proactive measures and strategies anchored on the three concepts and doctrines of D-PAD.

Collaboration, effective communication, and coordination among these many possible participants and stakeholders are essential to achieving project objectives, effectively managing resources, and delivering successful project outcomes. However, involvement of many participants and stakeholders creates complexities, interdependence, and navigation problems that can potentially trigger conflicts and disputes in the following ways:

1) *Challenging coordination*: Coordination can be challenging with the many project participants and stakeholders, and disputes may arise over scheduling and task assignments, resource allocation, and assumption of responsibilities. For instance, delays by a subcontractor or supplier can upend the entire project timeline and lead to disagreements and disputes over accountability for consequential liabilities and damages.
2) *Communication problems*: Communication breakdown is commonplace where there are this many project participants and stakeholders with different roles, perspectives, backgrounds, and experiences. This breakdown can lead to miscommunications, misunderstandings, errors, and delays, all of which can cause conflicts and disputes when project progress and outcomes are affected.
3) *Conflicting and competing interests and priorities*: With the many project participants and stakeholders, conflicting and competing interests and priorities can be expected. Discordant and unaligned interests and priorities are ripe grounds for conflicts and disputes. For instance, cost-saving alternatives proposed by the project owner may be at variance with the designed standards by the architect, designer, or engineer, leading to disputes over specification and scope changes.
4) *Vicarious responsibility can arise in contractual disputes*: While the privity of contract is between the project owner and the contractor, or between the contractor and the subcontractor, each party is still bound by the acts, omissions, commissions, or liabilities of their employees or subordinates. Such complex relationships and guilt by association can bring a lot of arguments, disagreements, and disputes over which party should take part or full responsibility for the negligent acts of employees or subordinates that led to liabilities and damages.
5) *Quality control and assurance difficulties*: With the many project participants and stakeholders, quality control and assurance are made all the more difficult. The difficulties compound with possible various interpretations and implementation of the various project aspects, leading to disputes over workmanship quality, adherence to standards, and compliance with regulatory requirements. Disputes arise over failure to meet expected agreed-upon standards and quality, potentially requiring costly corrections, amendments, and rework.

Preventing or avoiding disputes from the many players on the one project requires a culture of proactive inventorying and appreciation of the many participants and stakeholders, deploying QA/QC/TQM, Six Sigma, and PDM principles and tested management strategies, enthroning effective communication and coordination protocols, and clearly delineating roles and responsibilities in the contracts. Experienced and seasoned PMs are crucial in taking the lead here in facilitating collaboration, fostering a cooperative project culture, and dealing with any issues promptly and swiftly. Culture-consciousness must be exercised in recognition of the multicultural backgrounds and experiences that make up the project participants and stakeholders. Of course, where conflicts and disputes portend to insist to arise, swift steps must be taken to deescalate them at no cost of time and resources to the project.

Please refer to Table 14.1 for the recommended PDM's D-PAD concept applicable in this regard.

7.7 Team-Playing

Team-playing is crucial in the delivery of successful construction projects. It is of utmost importance due to the many project participants and stakeholders who need to work together toward cooperating to achieve collective project goals. Here are some ways team-playing manifests in project construction:

1) *Collaborative culture*: There is a recognition of the different cultures, backgrounds, and experiences, but efforts are nonetheless made to foster a collaborative atmosphere where project participants and stakeholders perceive themselves as a part of an interdependent and unified team or entity rather than individual or independent operators. Such an atmosphere encourages collaboration, sharing of resources and know-how, and a shared objective to complete the project on time, within budget, and to the agreed-upon quality and standards.
2) *Role and responsibility awareness*: Team-playing among the many project participants and stakeholders calls for awareness of, and clarity on, the delineation of roles and responsibilities of each member. Everyone understands their contributions to the successful delivery of the project. Each member recognizes that they bring specialized skills and expertise to the project.
3) *Open and timely communication*: Open and timely communication is a hallmark of an effective team. Effective communication coalesces the disparate team members, ensuring that everyone is up to date on the project progress, challenges, milestones, and any decisions, facilitating coordinated actions and timely adjustments as needed. To be effective, open and timely communication must be precise and succinct, honest, and freely and frequently exchanged among team members.

4) *Shared goals and objectives*: Team-playing promotes alignment of common project goals and objectives, which are primarily timely delivery, within budget, and to the designed quality standards. By promoting shared goals and objectives, project participants and stakeholders prioritize collective success over individual interests, thus fostering a cohesive team and collaboration.
5) *Participatory decision-making and problem-solving*: Team-playing encourages collaborative decision-making and problem-solving among project participants and stakeholders. By pooling perspectives and expertise, teams can identify and deploy creative and innovative solutions and mitigate risks and disruptions to the progress of the project.
6) *Trust and respect*: Team-playing promotes trust and respect among project participants and stakeholders. All the perspectives and expertise inherent in the team members are trusted and respected, allowing them to ventilate and contribute to the success of the project. Team members feel acknowledged, valued, and motivated to contribute and perform their best. Trust and respect among team members build confidence in collaborative efforts and promote open and transparent communication.
7) *Continuous improvement*: Team-playing enthrones an atmosphere of continuous improvement and learning from the successes and challenges of the project. Project participants and stakeholders actively and continually seek avenues and opportunities to enhance the processes, adopt and deploy best practices, and innovate to optimize project outcomes, thus further strengthening the team's cohesion and effectiveness.

Essentially, team-playing is pivotal in navigating the reality and complexity of modern projects and the many players on them, fostering synergy among diverse participants and stakeholders, and achieving successful project outcomes through collaboration, effective communication, and shared responsibility and accountability. Teams can leverage collective strengths and expertise to deliver successful projects that meet the owner's expectations and industry standards. While team-playing is essential and beneficial to fostering collaborative efforts to deliver successful projects, it can also be a trigger for disputes under certain circumstances:

1) *Bureaucracy and decision-making delays*: In a collaborative team setting, project participants and stakeholders are likely to prioritize consensus-building and collective decision-making over speed and efficiency; delays may arise from efforts to align perspectives or accommodate varying perspectives, which can impact the project timeline with consequential liabilities and damages that are triggers of disputes.
2) *Divergent goals and priorities*: While team-playing encourages alignment around shared project goals, participants and stakeholders may have differing

or divergent priorities that would color their views and interpretations of the project objectives and strategic direction. Disputes can be triggered in this situation, especially when one party perceives that another's decisions and actions are counter to the project's overall objectives.

3) *Assumption of responsibility and accountability*: With team-playing, there is a collective sense of responsibility for the project outcomes among participants and stakeholders. However, disputes can arise if there are disagreements over individual roles, responsibilities, and accountabilities for specific tasks and decisions. Lack of clarity in role delineation and assignment can lead to fault-finding, finger-pointing, and blame games if and when even minor challenges and issues arise that lead to costly, consequential quality issues, such as errors, omissions, commissions, or rework.
4) *Communication challenges*: Communication breakdowns can occur due to lingo differences, language barriers, communication styles and nuances, or misunderstandings. Communication breakdown or miscommunication can bring about confusion about expectations, project requirements, instructions, and directives, potentially leading to disputes over compliance with agreed-upon or contractual quality standards and outcomes.
5) *Resource and territorial fights*: Collaborative projects call for efficient resource allocation among project participants and stakeholders, including labor, materials, equipment, and operational space. Disputes can arise if there is a perception of inequitable resource allocation, skewed procurement decisions, and off-keel scheduling priorities. Project participants and stakeholders compete for the limited resources, each trying to outsmart the other, causing tensions and disputes as a result over prioritization and fair allocation or distribution.
6) *Cultural and organizational differences*: With diverse project participants and stakeholders coming together as a team, cultural and/or organizational differences in norms, practices, values, and perspectives can inject chaos into the collaborative dynamics. These differences may bring about misunderstandings and clashes that can impact effective communication, collaborative decision-making, and problem-solving, potentially leading to disputes over perceived disregard, biases, or unfair treatment.
7) *Challenges handling difficulties*: Differing viewpoints and personalities in a collaborative team are potential triggers of disputes in their midst during challenges and issues due to resultant frictions and tensions, clashes, and differing approaches to decision-making and problem-solving.

Preventing or avoiding disputes from team-playing requires a culture of appreciation of all project participants and stakeholders with their perspectives and idiosyncrasies, deploying QA/QC/TQM, Six Sigma, and PDM principles, and tested project management strategies and guidelines as propagated by Project Management Body of Knowledge (PMBOK), clearly establishing and allocating roles, responsibilities, and expectations for all team members, fostering open and transparent communication and coordination protocols to ensure that information is promptly and accurately shared among members, and promoting a culture of trust, respect, and mutual understanding among diverse team members to moderate, mitigate, or eliminate frictions from cultural and organizational differences. Experienced and seasoned PMs are crucial in taking the lead here in facilitating collaboration of diverse teams, fostering a cooperative project culture and shared project goals, and dousing any challenges and issues promptly and swiftly before they escalate to disputes requiring expenditure of resources and time to negotiate, mediate, or arbitrate.

Please refer to Table 14.1 for the recommended PDM's D-PAD concept applicable in this regard.

Chapter Review Questions

1) Attitudes and dispositions of project participants have no impact on construction disputes. True/False?
2) Communication difficulties among stakeholders can lead to disputes in construction projects. True/False?
3) Cultural disposition to work refers to the collective norms and behaviors that influence how work is approached. True/False?
4) Litigious participants are less likely to escalate minor issues into formal disputes. True/False?
5) Team-playing among stakeholders helps reduce the likelihood of disputes. True/False?
6) Cultural disposition to contracts has no effect on how parties interpret contractual obligations. True/False?
7) Having many players on a single project can increase the complexity and potential for disputes. True/False?
8) Differing interests and priorities among stakeholders can lead to misaligned expectations and disputes. True/False?
9) Which of the following is a key factor contributing to disputes due to attitude and disposition?
   a) Clear communication
   b) Differing interests and priorities
   c) Shared goals
   d) Defined roles
10) What is a common result of poor team-playing in construction projects?
    a) Increased collaboration
    b) Reduced costs
    c) Delays and disputes
    d) Improved quality

11) Which of the following best describes the impact of cultural disposition on contracts?
    a) It ensures all parties agree on the same terms.
    b) It influences how contracts are perceived and enforced.
    c) It eliminates the need for written agreements.
    d) It has no impact on dispute resolution.
12) Which of the following is NOT a flashpoint discussed in Chapter 7?
    a) Litigious nature of participants
    b) Cultural disposition to work
    c) Payment delays
    d) Team-playing
13) What is a recommended strategy to reduce disputes caused by miscommunication?
    a) Avoid meetings
    b) Use only verbal agreements
    c) Promote open and timely communication
    d) Limit stakeholder involvement
14) What is a common effect of cultural differences among project participants?
    a) Faster project delivery
    b) Improved budgeting
    c) Miscommunication and misunderstandings
    d) Reduced need for documentation
15) The tendency of some stakeholders to resort quickly to legal action is referred to as being ________.
16) A lack of ________ among project participants can lead to misunderstandings and disputes.
17) Cultural ________ to work affects how project participants approach tasks and responsibilities.
18) When too many stakeholders are involved in a project, it can lead to coordination challenges and increased ________.
19) Effective ________ among team members is essential for minimizing conflicts and achieving project goals.
20) A stakeholder's tendency to prioritize their own interests over collective goals can undermine ________ efforts.

Exercises

Exercise 7.1 Stakeholder Attitude Mapping

Objective:

Identify and assess the attitudes and dispositions of stakeholders in a real or hypothetical construction project.

Instructions:

- Choose a real or fictional construction project involving multiple stakeholders (e.g., owner, contractor, subcontractors, designers).
- Create a stakeholder map listing each participant.
- For each stakeholder, assess:
  - Their communication style (e.g., direct, indirect, formal, informal).
  - Their approach to conflict (e.g., collaborative, litigious).
  - Their cultural disposition to work and contracts.
- Identify potential flashpoints where attitudes and dispositions may clash.
- Propose strategies to prevent or douse disputes based on your analysis.

Outcome: Develop awareness of interpersonal dynamics and proactive dispute management strategies.

Exercise 7.2 Team-Playing Simulation

Objective:
Experience the impact of team-playing and miscommunication on project outcomes.

Instructions:

- Form small groups to simulate a project team (e.g., owner, contractor, architect, subcontractor).
- Each member receives a role card with a brief personality profile (e.g., litigious contractor, collaborative architect, rigid owner).
- The team must plan a project kickoff meeting and resolve a fictional issue (e.g., a change in scope or schedule).
- After the simulation, reflect on:
  - What attitudes and dispositions influenced the discussion?
  - What communication challenges arose?
  - What strategies helped or hindered resolution?

Outcome: Practice navigating interpersonal challenges and applying PDM principles in real time.

Discussion

1) How do cultural differences influence the way contracts are interpreted and enforced in construction projects? Share examples of how misunderstandings due to cultural disposition have impacted project outcomes.
2) Construction projects often involve numerous stakeholders. What are the risks and benefits of having many players on a single project? How can effective communication and team-playing reduce the likelihood of disputes?

8

Nature and Uniqueness of Project Dispute Flashpoints

Readers' Learning Objectives and Outcomes

1) *Analyze the various flashpoints and triggers of conflicts and disputes in the industry stemming from the nature and uniqueness of the project.*
2) *Conceptualize plausible prevention, avoidance, or dousing approaches that may be applicable to these conflicts' and disputes' flashpoints and triggers.*
3) *Match and apply prevention, avoidance, or dousing strategies effective or suitable to each conflict's and dispute's flashpoints and triggers.*

8.1 Introduction

The nature and uniqueness of a construction project refer to its distinctive architectural characteristics, the rarity of its features, its scope and requirements, together with its challenges that set it apart from other projects. Understanding this aspect of the project is crucial to mitigating associated risks to effectively plan, manage, and execute it. The broader constitution of the nature and uniqueness of a construction project includes:

1) *Architectural characteristics and features*: Defining the architectural characteristics and features of a construction project are dependent on the architect or designer engaged, the architectural aesthetics and style proposed by the designer, and the design requirements. For instance, for a museum project with a defined program and requirements, engaging Frank Gehry or Zaha Hadid and Salman Rushdie will produce different expectations and results in terms of architectural characteristics and features, often with vast cost differences. The contractor and all the project participants and stakeholders would have to sit up more to deal with the twists and turns of the features likely to be present in Frank Gehry's or Zaha Hadid's design. The situation would be different with designers who prefer less twisted design features. Remember, "Twisted buildings torture their builders!" Architects and designers follow different design philosophies. Some follow Louis Sullivan's "form follows function," while others go with "function follows form." By "form follows function," Louis Sullivan (1856–1924) articulated a design principle that proposes that the form or shape of a building or an object should directly relate to, and be dictated by, its intended function or purpose. Either school of thought comes with a big difference. The architects and designers' choice of materials can differ. Gehry uses a lot of titanium in his designs, and it can be a challenge for a contractor who has not worked with it. Overall, the design complexity can impact construction methods, material availability, and regulatory compliance as much as project timeline, cost, and quality. There may be more requests for information (RFIs) and changes, as well as rework from design complexity.

2) *Scope*: The nature and uniqueness of a construction project can dictate its scope, where the scope may be greater with more complex or unique designs in terms of the project's technical requirements, planning, construction strategies, and time, as well as resource allocation.
3) *Innovation and technical requirements*: Unique projects may incorporate unique building materials and advanced technologies that call for specialized engineering and innovative construction methods. Innovative technologies, such as building information modeling (BIM), sustainable building solutions, and modular construction, are increasingly embraced. They demand expertise and careful planning to integrate and implement for successful project delivery.
4) *Location of the project*: The nature and uniqueness of a construction project interact with its location and environment and influence its planning and execution. An innovative, unique project in an environment may be challenged by the geological constraints and weather conditions at that location; hence, a tailored solution may be needed to get it right. For instance, the challenges of Frank Gehry's $300 million Strata Center at Massachusetts Institute of Technology (MIT) ended in a lawsuit claiming that the building's unique forms and angles resulted in pervasive leaks, cracks, and drainage problems requiring costly do-overs.
5) *Risk profile and management*: Innovative unique projects come with higher risk profiles in terms of technical requirements, procurement, cost, planning and scheduling, as well as regulatory compliance. In other words, the risk factor is higher with more complex or unique projects.

In general, the nature and uniqueness of construction projects underscore their complexity and distinctiveness, as well as specific requirements in planning and successful execution. Recognizing and working to accommodate these realities early in the project process facilitates its success, minimizes risks, and maximizes participants' and stakeholders' satisfaction with solutions that meet the project goals and expectations. The key aspects of the nature and uniqueness of construction projects are further discussed in a greater detail under locational disparities and site conditions, novelty, RFI, detailing, and unforeseen circumstances.

8.2 Locational Disparities and Site Conditions

Locational disparities and site conditions play crucial roles in the delivery of construction projects, influencing everything from design and planning considerations to logistical challenges, labor, material, and equipment availability. Within are:

1) *Geographical and geological factors*: Geographical and geological factors can be vastly different from one region to another within a country. Some regions may

have seismic activities that would require appropriate design and construction standards prescribed by codes. Some may predominantly have clay soils, determining the foundation types and requiring measures to shore up building foundations to minimize or prevent destructive uneven settlement.

2) *Topography*: Topography and terrain can vary from one location to another. More hilly locations may present greater challenges and costs in excavation, foundation construction, and stabilization. Flatter terrains may present drainage problems, while rocky topography would call for specialized construction techniques.
3) *Climate and weather conditions*: Climate and weather conditions can vary from one region to another. Some may be tropical or temperate, with extreme heat prevalent in the former, and extreme cold in the latter. Some locations may have heavy snow or rainfall, while others may experience a drought-like or desert-like atmosphere. All these will shape project planning and design, the resources to be expended, and the challenges to overcome. Extreme hot or cold weather would require extensive insulation of the building envelope, external ducts, and pipes. Heavy rainfall areas would require enhanced drainage and flood control systems.
4) *Specific environmental regulations*: Local environmental regulations take precedence in dictating permissible construction methods and materials. Locations close to the coast or water bodies have stringent rules on siltation and sedimentation to protect and preserve aquatic life, which can influence the design, planning, management, and execution of building projects. The same applies to locations requiring strict maintenance of ecological equilibrium to protect flora and fauna. Hence, due to the location's ecosystems, such as wetlands or protected habitats, uncommon restrictions may be imposed on site development that would impact project design and construction activities to minimize environmental damage or impact.
5) *Market conditions and access to resources*: Market conditions vary regionally in terms of access to, and cost of, materials, labor, and equipment, which can impact project feasibility and budget. Remote locations are likely to experience greater challenges in sourcing skilled labor and transporting heavy equipment and materials, all of which can impact project cost and timeline.
6) *Infrastructure and utilities*: Locations and sites lacking adequate critical infrastructure and utilities—such as public water and sewer mains, electricity, and gas—may require costly upgrades or alternative solutions impact project cost and timeline.
7) *Cultural and historical factors*: Locations and sites of cultural or historical significance may require certain assessments and preservation requirements impacting project design and planning, as well as its cost and timeline.

As previously stated, the nature and uniqueness of a construction project interact with its location and environment and influence its planning and execution. Hence, understanding, respecting, and addressing locational disparities and site conditions are essential for successful project outcomes. By considering location and site-specific conditions in project planning and execution, participants and stakeholders can optimize resource allocation, mitigate risks, and achieve successful outcomes tailored to the nature and uniqueness of that project. Due to the significant impact of locational disparities and site conditions on project planning and execution, they can be underlying triggers of disputes in many ways:

1) *Climate and geographical disparities*: A project designed to be built in a seismic location may face seismic uncertainties that result in extra delays and costs, both of which are frequent triggers of disputes. Heavy rain and storms, as well as snow, can lead to similar dispute possibilities. Parties may disagree and fight over appropriate or cost- and time-efficient approaches to deal with unexpected conditions.
2) *Environmental regulations, compliance, and enforcement*: Disputes can arise if enforcement and compliance with site-specific environmental regulations result in delays and extra costs. A hospital extension project in Ann Arbor, Michigan, designed for a pile foundation, ended up costing much more when the noise ordinance forced a change from pile to caisson foundation due to the noise from piledriving that was beyond permissible levels around the hospital zone.
3) *Dispute from terrain and topography*: Disputes can arise if the complex, unique project calls for complicated excavations, foundation construction, and access for heavy construction equipment that would stretch the timeline and add extra costs due to the challenging terrain and topography. Unexpected geological conditions may require costly adaptations and cause delays.
4) *Soil and underground conditions*: Variations in soil types (clay, sand, silt, loam, etc.) and groundwater levels may introduce uncertainties and unexpected conditions for foundation stability and drainage, both of which can potentially disrupt and delay the project and cost more to deal with. Additional foundation, foundation shoring, or rework may be needed. This situation can bring about disputes from delays, increased scope, and extra costs, and the equitable or inequitable allocation of responsibilities and liabilities between the parties. Disputes from foundation settlement issues may be anchored on arguments over whether the soil was well compacted and consolidated or not.
5) *Infrastructure and utilities availability*: Disputes can arise from a lack of, or inadequate availability of, necessary infrastructure and utilities, if unexpected costs and extra works arise from alternative solutions or upgrades, as this would cause delays and creep the scope, affecting the project timeline and costs.

6) *Challenges from access to resources*: Locational or regional variations in labor, material, and equipment availability and costs affect project timelines and costs. Complex and unique projects may be more cost-effectively and time-efficiently built in certain regional locations than in others. There may be higher labor skill levels for a particular project in a particular region than in others.

Preventing or avoiding disputes from locational disparities and site conditions of a construction project requires a culture of strong and tested project management strategies and guidelines, as propagated by the Project Management Body of Knowledge (PMBOK) that takes a holistic and global rather than a narrow view of projects, enabling the acknowledgement and consideration of possible disparate locational and site realities of any project. Such strategies are comprised of comprehensive feasibility studies and site investigations and assessments to identify and manage potential challenges, robust risk assessment and management practices with adequate contingency and insurance coverage, collaborative participant and stakeholder engagement, and clear contractual provisions for unforeseen circumstances and subsequent risk allocations and responsibilities. Any challenges and issues should be swiftly doused before they escalate to resource-consuming resolutions.

Please refer to Table 14.1 for the recommended PDM's D-PAD concept applicable in this regard.

8.3 Novelty of Project

The novelty of a construction project refers to its distinction from existing counterparts in terms of its unique aspects, innovations, and distinctive features. Several aspects contribute to the novelty of a construction project, and they include:

1) *Architectural style and design innovation*: Earlier, an example was given of a museum project designed by Frank Gehry or Zaha Hadid and Salman Rushdie. These designers' projects are uncommon and come with distinct, iconic architectural styles and design innovations for which they are noted. Some of their designed projects come with unique forms and structures that challenge conventional norms and introduce new concepts. Green building and sustainability considerations can introduce innovations that are novel.
2) *Technological integrations*: Deploying cutting-edge technologies can contribute to the novelty of a project in its design, planning, and execution. For instance, the use of BIM for 3D and 4D modeling and clash detection, and the application of artificial intelligence or augmented intelligence (AI) as per Ilozor (2024), machine learning, virtual reality (VR), augmented reality (AR), and mixed

reality (MR) for animation and simulation, are game-changing for optimizing project design and construction process speed and efficiency. Robotization will also transform the processes extraordinarily. Prefabrication and modular construction shift most of the processes off-site and into controlled conditions, and this can introduce novelty. Another novelty is the increased integration of smart building systems in projects such as the Internet of Things (IoT) devices, sensors, actuators, and other automations, enhancing building operation and management, energy efficiency, and occupant comfort.

3) *Innovative materials and construction techniques*: The use of innovative materials and construction techniques can differentiate a building project from others and introduce an attribute of novelty. Innovative materials include building with carbon fiber, reinforced cement fiber, quick-curing concrete, lightweight concrete, insulating concrete akin to insulated concrete forms, and alternative wood from recycled construction waste. Innovative construction techniques include 3D printing of whole buildings, robotic construction, advanced formwork systems, and packaged modular construction.
4) *Environmental and sustainability considerations*: Some environmental and sustainability considerations can make a building project stand out as novel. Uncommon integrations such as net-zero energy building (consuming as much energy as generated from or by the building) and zero carbon building (emitting zero carbon into the atmosphere from building materials to operation) can qualify a building project as novel, as are various passive design approaches, such as natural cooling, heating and lighting, and borrowing natural light to minimize energy consumption from both benign and nonbenign sources.
5) *Project scale and complexity*: A construction project can be novel from the sheer scale and enormity of its scope and complexity. Megaprojects are uncommon projects that often innovate as new ones are being built, e.g., airports, seaports, stadiums, refineries, dams, levees, and some urban infrastructure projects. From shell and harsh concrete construction to the use of a vast network of space frames, airport construction has continued to innovate ever so infrequently as airports are being built around the world.
6) *Culturally and historically significant projects*. Culturally and historically significant projects can have attributes of novelty where clever combinations of ancient and modern features marry seamlessly to the admiration and satisfaction of current users. Clever adaptive reuse and restoration are often observed with these projects, embodying a combination of astute preservation and modern functionality and systems.

By pushing the boundaries, construction projects can integrate new ideas to address emerging challenges and provide maximal outcomes and satisfaction for stakeholders. However, the novelty of a construction project can trigger disputes due to the unique challenges and complexities involved with it, and some of these sources include:

1) *Dispute from technical complexity and innovation*: Projects being built with new construction technologies and innovations often tend to encounter unforeseen technical challenges and performance issues. Disputes may arise if the project participants and stakeholders disagree on how best to address or navigate around the challenges or performance issues. Delays or quality issues from technical glitches or incompatibility are triggers of disputes.
2) *Dispute from new materials and construction techniques*: Projects using novel materials and techniques may face challenges of availability and supply chain, installation know-how, performance issues, or compliance with regulatory standards. Disputes can arise over supply chain delays and the suitability or quality of the materials, especially if they are perceived not to meet expected performance criteria.
3) *Unwieldy RFI, changes, and scope creep*: Construction projects adjudged as novel in design and construction often experience way more RFIs and changes from complexity and newness, leading to changed scope and scope creep that may take extra time and cost to address. Oftentimes, dealing with the sheer enormity of the twists and turns from multifarious angles and curves can be quite daunting and exasperating. The frequent iterations and modifications during the construction process take time and money that can burst the budget and extend the project timeline, both of which are potential triggers of disputes between the parties.
4) *Environmental and regulatory compliance*: Projects incorporating environmental and green features for higher sustainability may face challenges associated with the requirements for compliance with environmental regulations and green building standards for certification. For instance, projects going for the United States Green Building Council's (USGBC) Leadership in Energy and Environmental Design (LEED) certification encounter challenges not faced by others not going for such a certification. Disputes can arise over the interpretation and implementation of the credits for certification, and extra resources are needed, which may stretch the project timeline and costs. LEED projects often make credit interpretation requests and undergo credit interpretation rulings, all of which take extra time.
5) *Incompatibility of contract with novelty*: Some project contracts may not have been prepared to handle the complexities and ambiguities from novelty, and this would lead to misinterpretation and misapplication that can trigger disputes between the parties.
6) *Participants and stakeholders' expectations and communication*: Project participants and stakeholders respond differently to novelty, with the associated

complexities and ambiguities. Discordant interpretations, expectations, and miscommunications are rife in this scenario, and a tense work atmosphere can be created, leading to potential conflicts and disputes on how to tackle the challenges and who takes responsibility for any delays and consequential costs and damages.

Preventing or avoiding disputes from the novelty of a construction project requires a culture of strong and tested project management strategies and guidelines grounded in the PMBOK that draw from the global experience of different types of projects (simple and complex), enabling comprehensive risk assessments, contractual alignment, and fitness, together with fostering needed collaboration and communication. Such strategies deploy extensive feasibility and buildability studies, site investigations, and assessments to identify and manage potential challenges; robust risk assessment and management practices with adequate contingency and insurance coverage; collaborative participant and stakeholder engagement; and clear contractual provisions for unforeseen circumstances and subsequent risk allocations and responsibilities. Any challenges and issues should be swiftly doused before they escalate to resource-consuming negotiations, mediation, arbitration, or litigation.

Please refer to Table 14.1 for the recommended PDM's D-PAD concept applicable in this regard.

8.4 Rate of Request for Information

Request for information (RFI) has been previously explained. However, it is discussed here again under "nature and uniqueness of project-related dispute flashpoints" due to its outsized impacts in this scenario. The rate of RFI refers to the frequency with which contract-impacting requests for clarifications, additional details, changes in design and specifications, or contractual obligations occur during a project's delivery process. RFIs play a crucial role in unraveling the challenges of unique projects to successfully deliver them. They enhance clarity and minimize or eliminate associated uncertainties that hinder project progress and quality. Here are some factors influencing the rate of RFI under this scenario:

1) *Project uniqueness and complexity*: The rate of RFI is higher with unique and complex projects due to factors such as their novelty, size, design intricacies, and technical requirements that may demand more clarifications, additional design, and specification details, as well as contractual obligations.
2) *Clarity and suitability of contracts*: As the existing suite of contracts for projects may not be complete or comprehensive enough for novel or complex projects, more than usual RFIs may occur in this type of project. The contract may not

quite suit or be adequate for a novel project, and RFIs will be used to continually adjust or amend it as the project progresses from start to finish. Novel projects demand continuous learning from participants and stakeholders in both contracts and the construction delivery process.

3) *Team experience and expertise*: The experience and expertise of participants and stakeholders can influence the rate of RFI, as many are unequally experienced or exposed to the novelty. Less-experienced participants and stakeholders will demand more RFIs as they navigate the intricacies and challenges of a novel project with which they may have limited expertise. Note that the novelty of a project to participants and stakeholders varies. That is, participants and stakeholders may have different expertise and be differently experienced in a particular novel project, with some possibly having no experience at all. It often happens that a project that is novel to a firm can be executed by that firm through a quick merger or a joint venture arrangement. That would mean having people on the team with little or no experience in this novel project.
4) *Location and site conditions*: Unforeseen site conditions almost always lead to RFIs. The rate of RFI increases with a novel project that calls for complicated excavations, foundation construction, and access for unique heavy construction equipment, depending on the location. More hilly locations may present greater challenges in excavation, foundation construction, and stabilization, needing RFI for clarifications, further design and specification details, or a change in work scope and contractual obligations. Flatter terrains may present drainage problems, while rocky topography would call for specialized construction techniques, all of which may demand RFIs to unravel or alleviate the challenges one way or another.
5) *Shifting owner expectations*: A unique project is likely also novel to the owner, who may have to fill in the details as it progresses in construction. As the owner progressively understands the novel project and requests adjustments or changes, the contractor formally handles the requests through the RFI process to make a change order that legitimizes the requests and changes relative to the contract. Hence, as the owner's shifting expectations are more in novel projects, so also are RFIs.
6) *Regulatory requirements*: The unique project may be novel to existing regulations; hence, aligning it to regulatory requirements may be a work in progress, and a higher-than-usual rate of RFI will be necessary in seeking guidance on specific requirements, compliance, or approvals needed to continue with construction activities.

In summary, the higher rate of RFIs in novel projects is a measure of their intricacies, uncertainties, and challenges. It reflects the fluid and dynamic nature of novel projects and the ongoing need to continuously clarify, coordinate, and adapt them to conditions and the learning curve. The rate of RFIs can elevate the dispute potential in a novel project in the following ways:

1) *Timing and delays*: The RFI process takes time, and the more the RFIs, the more the time they take. Such extra time taken feeds into delays that can stretch the project timeline. For instance, the then and now novel Sydney Opera House project was supposed to be completed and commissioned in 4 years, by January 1963, but it was completed 10 years behind the scheduled timeline and 14 years after construction started. There's no gainsaying that it was disputatious where they had to chase away the architect, Jorn Utzon, and only invited him back to visit his iconic project in 2023, a whole 60 years after Sydney Opera House was completed. The Sydney Opera House's novelty presented unique structural engineering difficulties. Its location by the seaside and the weather presented their challenges and led to regular work stoppages that significantly delayed progress and stretched the project timeline. Due to its subsequent long delivery timeframe, there was a change in government, and each demanded and made design changes. Public opinion turned against the project, and Jorn Utzon, who was angry about the changes, was run out of Australia back to his native Denmark.
2) *Changes and scope creep*: Due to the many changes and RFIs prevalent in novel projects, scope creep can be expected, and disputes can arise over contests about the necessity of the changes and their extent. Jorn Utzon was angry over the necessity of the changes that altered or violated the sanctity of his original concept, and the contractor was not happy either, but the job must be done, and someone bore most or all of the cost and other implications.
3) *Significantly altered contract*: RFIs that result in change orders amend or alter the contract to some extent. Hence, the more the RFIs, the more the change orders, and the more the contract is altered from the original. Contract alterations and their broad implications are potential triggers of disputes between the parties over disagreements about apportioning of associated responsibilities and liabilities.
4) *Document handling and control issues*: Due to the many RFIs in novel projects, documenting and controlling them can be daunting, and this may lead to confusion and omissions that can be contested and disputed.
5) *Cost implications*: Novel projects have contentious and disputatious cost implications due to their already-discussed attributes. Again, Sydney Opera House was budgeted to be completed with just $7 million, but it ended up costing $102 million, nearly fifteen times its original budget. You can imagine the extent of disputes associated with this astronomical budget burst.
6) *Quality and performance expectation issues*: The quality of a project is measured by the extent to which it meets the design and designed intents. If the

outcome deviates significantly from design and designed intents, quality and performance expectations are affected, and these are potential triggers of disputes between the parties. The deviations are accomplished with RFIs, and the more the RFIs, the more the deviations, together with outcome quality and performance relative to design and designed intents.

In order to prevent or avoid disputes from a high rate of RFI on novel construction projects, you must imbibe a culture of the PMBOK that draws from global experiences of different types of complex and megaprojects, enabling comprehensive risk assessments, contractual alignment, and fitness, together with fostering needed collaborations and communications. Such a culture deploys extensive feasibility and buildability studies way ahead of construction start, site investigations and assessments to identify and manage potential challenges, robust risk assessment and management practices with adequate contingency and insurance coverages, collaborative participant and stakeholder engagement, and clear contractual provisions for unforeseen circumstances and subsequent risk allocations and responsibilities. A precontract addenda process should be exhaustively utilized in order to minimize or significantly eliminate RFIs as the project construction progresses. Any challenges and issues should be swiftly doused before they escalate to disputes that call for resource-consuming negotiations, mediation, arbitration, or litigation.

Please refer to Table 14.1 for the recommended PDM's D-PAD concept applicable in this regard.

8.5 Poor Detailing of Novel Structures

Unique or innovative architectural features are, sometimes, relatively new to their designers. Hence, some suffer from inadequate or poor detailing and specification of their materials and construction methods, as their initiators or originators grapple with them through their learning curve. Some of the causes of poor detailing here include:

1) *Novelty of design features*: The novelty of the project's design features means that the design may be new to them all, and they may have limited or no experience and expertise on them, as they are being introduced for the first time. This leads to a lot of trial and error, both at the project design and construction stages. The intricate architectural forms and innovative engineering solutions of novel projects may exceed the capabilities or experience and expertise of the design teams, but they don't disclose this fact to other stakeholders while muddling with them. This scenario makes it impossible for the inadequate or limited-experience design team to anticipate all the

challenges to be encountered during construction and comprehensively detail all the intricacies in their designs and specifications, a priori or a posteriori.

2) *Ambiguities and inconsistencies in documentation*: The novel projects present significant ambivalence and ambiguity to their designers. The owner may hire a particular designer for what it expects to achieve by so doing, which may be either functional or aesthetic, or both. The owners trust that the designers know what they are doing, but only the designers are aware of the limits of their true capabilities to give the owners what they expect from them without many problems. As has been alluded to, the ambiguities and inconsistencies that can be witnessed in a novel project call for many back-and-forth steps in working with detailed designs and specifications, which end up more or less half-baked or inconclusive.
3) *Imposition of alien structure on a location or site*: Most novel projects are an imposition on their locations or sites. As such, they are forced to align with the locational or site realities, which may be incongruent. The novel project may be suited to a flat site and detailed as such in its design and specification, and if forced or sited on a hilly location, it can experience unforeseen geotechnical challenges that might have been adequately detailed in its design and specification. Locational and environmental conditions, such as seismic activity, high wind, and heavy snow, may not support a novel project imposition, and its design and specification may poorly or inadequately anticipate, reflect, and accommodate these realities.
4) *Material and construction compatibility*: Rarity and little familiarity with appropriate construction techniques of novel project features and structures can result in errors in detailing unique materials and features assembly or installation, and this may jeopardize their proper construction with intact functional and structural integrity.

Overall, poor detailing of novel structures is injurious to their delivery and to the participants and stakeholders, potentially launching parties into disputes emanating from it. Here are some of the associated disputes from poor detailing of novel structures:

1) *Design errors and misinterpretations*: Design errors occasioned by a project's novelty with scanty design and specification details are open to interpretations and misinterpretations, and when that leads to incorrect construction, corrections, and rework that takes more time and costs more, disputes can be triggered between the parties as to the apportioning of responsibilities and consequential liabilities. For instance, if drawings and specifications lack adequate details on the assembly and installation of innovative structural elements and features, contractors can assemble and install them incorrectly, and they

will be compelled to correct or rework them, leading to disputes over apportioning of responsibilities and consequential liabilities for the errors.

2) *Half-baked designs and specifications*: Where novel project designs and specifications are insufficiently detailed, ambiguities and inconsistencies may be profuse, causing self-help and confusion during construction. Contractors face the challenge of determining the correct procedures or requirements to build correctly, and this scenario can trigger tensions, conflicts, and disputes between the parties over conformance with the design and specifications.
3) *Technical challenges from unforeseen site conditions*: Novel projects come with uncommon technical challenges from their complexities and intricacies that don't jell well with unforeseen site conditions. If these connections are not fully thought about and detailed and worked out completely ahead of construction start, many RFIs and changes would follow, impacting the project timeline and costs, which are ready dispute triggers. The novel project's compatibility with the existing infrastructure of a location and its unforeseen geological realities can trigger disputes between the parties over extra costs and delays.
4) *Quality and performance issues*: The quality of a project has been discussed as measured by the extent to which the outcome met the design and designed intents, namely, functional and aesthetic. Poor detailing of a novel project's documentation may result in deviations from this metric, and disputes can arise over the outcome not meeting quality and performance standard expectations. For example, if poor detailing leads to incorrect or improper installation of insulation in novel projects, and such failure leads to the outcome not meeting energy efficiency standards expected, disputes can be triggered.
5) *Contractual and liability issues*: Due to the novelty of the project, the contract used for its construction may not sufficiently cover all aspects and intricacies of the associated challenges and responsibilities, and this scenario can lead to arguments and disputes over the apportionment of responsibilities and associated consequential liabilities. In other words, the contract may be inadequately detailed, as are designs and specifications, due to the novelty of the project.
6) *Schedule implications*: Poor detailing of novel projects and dealing with their fallout take time to resolve, and this can stretch the timeline, leading to disputes from delays and consequential damages.
7) *Cost implications:* Poor detailing of novelty projects and dealing with their fallout costs extra to resolve, and this can stretch or burst the budget, leading to disputes over apportioning of responsibilities and consequential damages.

Preventing or avoiding disputes from poor detailing of novel construction projects requires a culture of quality assurance (QA), quality control (QC), total quality management (TQM), Six Sigma, and PDM with attention to detail deployed every time, including when handling complex and megaprojects. This culture calls for enabling comprehensive risk assessments and management, contractual alignment, and reviews for fitness, together with fostering needed collaboration and communication as the project progresses. Such a culture deploys extensive feasibility and technical buildability studies with comprehensive preconstruction document reviews, site investigations, and assessments to identify and manage potential inadequacies and challenges; robust risk assessment and management practices with adequate contingency and insurance coverage; collaborative participant and stakeholder engagement; and clear contractual provisions for unforeseen circumstances and subsequent risk allocations and responsibilities. A precontract addenda process should be exhaustively utilized in order to minimize or significantly eliminate or clarify poor details before construction starts. Any challenges and issues should be swiftly doused before they escalate to disputes that call for resource-consuming negotiations, mediation, arbitration, or litigation.

Please refer to Table 14.1 for the recommended PDM's D-PAD concept applicable in this regard.

8.6 Unforeseen Circumstances

Unforeseen circumstances refer to unexpected conditions or situations that arise in the course of building a novelty project, but are not unanticipated and prepared for at the design and planning stages, a possibility that tends to be more prevalent in this type of project that impacts its feasibility, cost, and schedule, with the associated implications on disputes. Some unforeseen circumstances that don't jell well with novel projects are beyond what has been discussed previously, and here is more about them:

1) *Geotechnical circumstances*: Certain geological issues may be unforeseen at a novel project's location and have a significant impact on its feasibility, cost, schedule, and outcome, especially as they have to do with its excavation and foundation. Geological realities include such things as the bearing capacity of the soil, the water table around the site, and the susceptibility of the land to erosion. Some unforeseen geological issues encountered can necessitate an in-between construction change of foundation design and its excavation, shoring, and stabilization.
2) *Environmental factors*: Some of the prevalent natural environmental conditions may be unforeseen. Flooding, extreme wind conditions, such as severe storms, hurricanes, tornadoes, and seismic activities, may not be accurately foreseen from year to year. They can be more injurious and consequential to the schedule, cost, and overall outcome of a novel project.
3) *Materials supply chain*: Depending on the project's location, the materials supply chain and price stability may not be accurately projected. Therefore,

extra lead time may have to be allowed for the procurement of materials and equipment.

4) *Availability of labor*: Depending on the project's location, labor availability and skills may not be accurately predictable. For instance, even if the type of labor that may be available is known, such as union and nonunion labor, it is hard to predict when union labor may embark on industrial actions and demand far higher compensation than has been estimated for the novel project already under construction. It is a problem if the contractor has to cut corners and recruit alternative labor of mixed skills under these circumstances.
5) *Infrastructure and utility availability and failure*: The availability and adequacy of existing site infrastructure are not fully known ahead of construction start. Roads and bridges to the site may fail for any reason without notice, as may electric power supply lines, telecommunications, water, sewer, and gas mains. The consequential disruptions can be vast.
6) *Regulatory requirements*: There may occur unforeseen changes to building codes, zoning, or environmental regulations during the course of the project. There may be a requirement for additional approvals and permits, design revisions, and compliance with measures and requirements that were not planned or budgeted for. This can be more complicated and consequential for novel projects.
7) *Contract documents' errors or inadequacies*: Errors, omissions, and commissions that become apparent during construction across contract documents (drawings, specifications, and contracts) can be more prevalent in novel projects due to the complexity, innovations, and intricacies. Such an occurrence will lead to design revisions for the changes, corrections, and rework it calls for.

Novel projects are hard to execute completely successfully, and unforeseen circumstances make it even harder. Unforeseen circumstances are frequent dispute triggers in novel project construction delivery, and here are some of the ways this happens:

1) *Design and specification issues*: Design and specification modifications or revisions resulting from unforeseen circumstances, which tend to be more impactful on novel projects, can trigger disputes over which party to attribute the errors, omissions, or commissions, and the changes and consequences arising from them. The necessity or relevance of the changes can be debatable and disputable.
2) *Schedule delays*: Unforeseen circumstances in novel projects and dealing with their fallout take time to resolve, and this can stretch the timeline, leading to disputes from delays and consequential damages. Weather conditions, regulatory, geological, and environmental surprises can occur unannounced and

disrupt the construction schedule. Disputes can come from requests for time extensions, which may be viewed as disproportionate or unreasonable, as may any claims of consequential damages.

3) *Cost overruns*: Unforeseen circumstances in novel projects and dealing with their fallout cost more to resolve, and this can stretch or burst the budget, leading to disputes over attribution and apportioning of responsibilities and consequential liabilities. Contractor claims may be denied on the grounds of the contested adequacy of contingency measures taken. For instance, unpredicted events of back-to-back hurricanes took down a construction crane and knocked down parts of the project, but the owner disputed and nearly wholly denied the contractor's claim for the extra costs to remedy the situation with the argument that the contractor failed to take adequate measures to minimize the damages from the hurricane.
4) *Contractual and liability issues*: As already stated, due to the novelty of the project, the contract used for its construction may not sufficiently cover all aspects and intricacies of the associated challenges and responsibilities, and this scenario can lead to arguments and disputes over apportioning of responsibilities and associated consequential liabilities. In other words, the contract may be inadequately detailed as to designs and specifications due to the novelty of the project. Hence, the contract becomes a work in progress that will continue to be adjusted or amended as project construction progresses.
5) *Regulatory compliance*: Unforeseen changes to building codes, zoning, safety, or environmental regulations during the course of a novel project may require additional approvals and permits, design revisions, remediation, and compliance with measures and requirements that were not planned or budgeted for. The parties can dispute over who is responsible for implementing the measures and the associated costs.
6) *Force majeure clauses interpretation*: Unpredictable force majeure events, such as natural disasters, pandemics, and political upheavals that are beyond the control of the parties, can trigger disputes in their contractual clauses' interpretation and application. COVID-19 brought vast disputes on projects across the world, as most contracts did not envisage such an occurrence, and it is left to arguments and disputations for the parties as to how to interpret and apply them with the consequential disruptions and damages. Disputes arose over whether the event qualified as force majeure (in the absence of its provision in the contract clauses), contractual obligations, and the extent of relief or compensation to be allowed for it, and which party to bear or earn part or all of it.

Preventing or avoiding disputes from unforeseen circumstances in novel construction projects calls for a culture of QA/QC/TQM, Six Sigma, and PDM with attention to detail deployed every time, including when handling complex and megaprojects. This culture calls for enabling comprehensive risk assessments and management, contractual alignment and review for fitness, together with fostering needed collaborations and communications as the project progresses. Such a culture deploys extensive feasibility studies and site analysis with comprehensive preconstruction contract review and alignment, site investigations and assessments to identify and manage potential unforeseen circumstances and possible challenges they can bring along, robust risk assessment and management practices with adequate contingency and insurance coverages, collaborative participant and stakeholder engagement, and clear contractual provisions for unforeseen circumstances and subsequent equitable risk allocations and responsibilities. Any challenges and issues should be swiftly doused before they escalate to disputes that call for resource-consuming negotiations, mediation, arbitration, or litigation.

Please refer to Table 14.1 for the recommended PDM's D-PAD concept applicable in this regard.

Reference

Ilozor, B.D. (2024). SpaceDEC as AI bot assisting design and building thinking for efficient space allocation and energy conservation. Presented at *AI and the Future of Design Conference*, April 20, 2024, Covenant University, Nigeria.

Chapter Review Questions

1) The nature and uniqueness of a construction project can influence the likelihood of disputes. True/False?
2) Architectural complexity has no impact on project risk or dispute potential. True/False?
3) Locational disparities, such as terrain and climate, can trigger construction disputes. True/False?
4) A high rate of requests for information (RFIs) is often a sign of clear and complete documentation. True/False?
5) Unforeseen circumstances can be eliminated through planning. True/False?
6) Poor detailing of novel structures can lead to errors, delays, and disputes. True/False?
7) Unique projects often require more collaboration and communication among stakeholders. True/False?
8) Which of the following is a common dispute flashpoint in unique construction projects?

a) Standardized design
b) Poor detailing of novel structures
c) Routine scheduling
d) Familiar materials

9) What is the major consequence of high RFI rates in a project?
a) Reduced costs
b) Improved communication
c) Project delays and scope creep
d) Elimination of disputes

10) Which factor is most likely to contribute to disputes in a novel project?
a) Use of standard materials
b) Clear contract terms
c) Innovative design features
d) Routine site conditions

11) Which of the following is NOT a dispute flashpoint discussed in Chapter 8?
a) Novelty of the project
b) Rate of RFIs
c) Payment delays
d) Poor detailing

12) What is a recommended strategy to reduce disputes in unique projects?
a) Avoid collaboration
b) Ignore site conditions
c) Conduct thorough feasibility studies
d) Use outdated design templates

13) Which of the following best describes the impact of unforeseen circumstances on construction projects?
a) They are predictable and easily managed.
b) They often lead to delays and increased costs.
c) They improve project efficiency.
d) They reduce the need for RFIs.

14) What is a key reason why novel structures may be poorly detailed?
a) Designers are always experienced with new forms.
b) Novel structures are simple to build.
c) Designers have limited prior experience with unique features.
d) They follow standard building codes.

15) The ________ of a project refers to its distinctive features and challenges that set it apart from others.

16) ________ conditions such as terrain and climate can significantly impact project planning and execution.

17) A high rate of ________ often indicates unclear documentation or design complexity.

18) ________ structures often suffer from poor detailing due to limited experience with their design.
19) ________ circumstances are unexpected events that can disrupt project timelines and increase costs.
20) Architectural ________ and features can increase the risk of disputes due to their complexity.

Exercises

Exercise 8.1 Site Condition Scenario Planning

Objective: Explore how locational disparities and site conditions can trigger disputes.

Instructions:
- Choose a real or hypothetical project site with challenging conditions (e.g., steep terrain, high water table, seismic zone).
- Create a scenario where unexpected site conditions are discovered after construction begins.

In small groups, answer:
- What contractual clauses should have addressed these conditions?
- How could the design team be better prepared?
- What dispute could arise, and how would you douse it using D-PAD strategies?

Outcome: Learn to anticipate and manage disputes arising from site-specific challenges.

Exercise 8.2 Novelty versus Standardization Trade-Off

Objective: Evaluate the risks and benefits of pursuing architectural uniqueness.

Instructions:
- Compare two projects: one with a highly innovative design (e.g., twisted façade, smart materials) and one with a conventional design.

Identify:
- Key differences in detailing, RFIs, and stakeholder coordination.
- Potential dispute triggers unique to the innovative project.
- Strategies to mitigate those risks without sacrificing design intent.

Outcome: Understand how innovation impacts dispute potential and project delivery.

Discussion

1) Is architectural innovation worth the increased risk of disputes in construction projects? How can teams balance creativity with constructability and risk management?
2) High RFI rates are common in complex projects. Do RFIs reflect poor documentation, or are they an inevitable part of innovation? How can teams reduce RFIs without stifling creativity?

9

External Factors-Related Dispute Flashpoints

Readers' Learning Objectives and Outcomes

1) *Analyze the various flashpoints and triggers of conflicts and disputes in the industry associated with external project factors.*
2) *Conceptualize plausible prevention, avoidance, or dousing approaches that may be applicable to these conflicts' and disputes' flashpoints and triggers.*
3) *Match and apply prevention, avoidance, or dousing strategies effective or suitable to each conflict's and dispute's flashpoints and triggers.*

9.1 Introduction

External factors in construction refer to those conditions and influences that are beyond the control of the project team, but still have a significant bearing on the planning, execution, and outcomes of the project. These factors originate from various sources with varying degrees of impact on the project process, and they include weather conditions, force majeure, regulatory enforcements, politico-economic changes, differential lead times, labor unions, and labor seasonality. These factors are discussed in further detail hereunder.

9.1.1 Weather Conditions

Weather conditions such as rain, wind, snow, ice, humidity, and heat have a significant impact on construction projects, most of which are currently predominantly accomplished outdoors in uncontrolled environments. They can impact project progress, cost, safety, and quality. Understanding and managing these conditions are crucial to maintaining timeliness and productivity while managing risks and safety. Specific weather conditions affect projects in the following ways:

1) *Rain or precipitation*: Excessive rainfall can cause soil erosion, waterlogging, and failure of excavation, shoring, and grading. It can affect concrete curing, and it can affect the quality of concrete and asphalt paving. Water management may be called for, including pumping out water from foundation trenches. While precipitation can be predicted on short terms, it is difficult to factor the predictions into the relatively longer-term projects.
2) *Extreme temperatures*: Extreme heat or cold can affect concrete curing, causing cracks and weakening its development of strength. Freezing temperatures can introduce harmful free-thaw situations in some materials that weaken their performance and strength, and ultimately their quality.
3) *Extreme windy conditions*: Extreme windy conditions can endanger the safety of workers, crane operations, and compromise the stability of temporary structures such as formwork, scaffolding, and wraps.
4) *Snow and ice*: In regions that witness snow and ice, both can frustrate construction by delaying excavation, material handling, and worker safety from hazardous conditions. Certain construction may have to wait under snowy or icy conditions, and others may need adjustments in their processes.
5) *Humidity*: High humidity can affect painting, plastering, and adhesive application by prolonging curing time and impacting quality. Moisture from high humidity can lead to the quick degradation of building materials and the development of mold and corrosion.
6) *Lightning and thunderstorms*: Both can delay or suspend some construction activities. Every year, many workers lose their lives to these weather conditions,

where they are not well managed and respected in the construction process, especially at heights and while working on metal structures.

Overall, extreme weather conditions can dictate the outcome of a project depending on how it is considered, handled, or managed. If not properly considered, handled, or managed, they can bring about disputes, especially with respect to their impact on increased costs, messed-up project timelines, and compromised quality. Here are some of the associated disputes from weather conditions:

1) *Project schedule delays*: Adverse weather conditions, such as heavy rainfall, thunderstorms, extreme windy conditions, and lightning, can cause work delays or stoppages. Contractors may seek extra time to complete construction, but the owners may dispute the severity and the due diligence of the contractors in managing and mitigating weather-caused delays.
2) *Force majeure disputes*: The contractor and the owner may dispute over what rises to force majeure as cited in the contract. In most contracts, force majeure exculpates the contractor from responsibility and liability over any consequential fallouts, but the owner may dispute that the contractor has not exerted due diligence to manage and mitigate the disruptions.
3) *Cost overruns*: Extended labor hours, equipment rental times, and mitigation measures all bring with them extra costs over and above the contract sum. Disputes may arise between the parties over what is legitimate to claim in extra costs, who is responsible or liable for what, and any contingency provisions.
4) *Quality of work and performance*: As has been stated, weather conditions can affect the quality and performance strength of concrete and asphalt. Compromised quality falls under a breach of contract and can be the subject of disputes between the parties.
5) *Safety concerns*: Weather conditions can bring about hazardous work conditions that can be injurious to workers, leading to delays, work stoppages, and claims. Disputes can arise over decisions on safety provisions, actions, and reactions due to severe weather conditions. Extracontractual demands may be made to provide or implement adequate safety measures by either or both parties.

It is more plausible to avoid disputes from weather conditions rather than to prevent them through having existing cultures and structures in place. Avoiding disputes from extreme weather conditions in construction projects calls for proactive planning, incorporating weather forecasting and monitoring into project planning and scheduling; clear and concise contractual provisions about weather-related risks, delay claims, and force majeure events prevalent at the project location; collaboration and equitable risk allocation and modalities in any weather-related events; and maintenance of clear, effective, and open communication to swiftly address and document weather conditions, work disruptions, and mitigation efforts applied. Any challenges and weather-related issues should be swiftly doused before they escalate to disputes that call for resource-consuming negotiations, mediation, arbitration, or litigation.

Please refer to Table 14.1 for the recommended PDM's D-PAD concept applicable in this regard.

9.1.2 Force Majeure

Force majeure refers to a legal concept acknowledging that certain occurrences are acts of God and beyond the control of both parties to a contract. These occurrences or factors make it impossible for the parties to fulfill their obligations to the contract. Force majeure has an outsized impact on construction projects, which is why it is often cited in many contract clauses, and it includes unforeseeable natural disasters, war breakout, strikes and boycotts, pandemics like COVID-19, government action, and any other events considered way beyond the reasonable control of the parties to the contract. Here are key elements of force majeure often encountered in construction contracts:

1) *Force majeure events that may be specified in the contract*: Common examples in construction contracts include earthquakes, hurricanes, floods, fires, epidemics or pandemics, and government actions, such as expropriation of funds or changes in the law.
2) *Force majeure consequences*: The inclusion of force majeure clauses in the contract can exculpate the contractor from the consequences of project delays and extra costs occasioned by the classified or named event. In other words, it excuses delays in the performance of contractual obligations impacted by the force majeure event. For example, due to a hurricane event, construction may be halted due to safety concerns and mitigation efforts, thus delaying the project timeline and milestones. Common provisions on the account of work disruptions from a force majeure event include time extensions, cost adjustments, and excuses from performance of impacted contractual obligations.
3) *Notification requirements*: Some contracts typically require that the affected party notify the other of the occurrence of a force majeure event, its impact on work performance, and the steps taken to mitigate or ameliorate its impacts.
4) *Mitigation obligation*: The party invoking force majeure is obligated by the contract to take reasonable steps to mitigate the impact of the event on work performance, which may include securing the site, reallocating resources, and implementing alternative, helpful workplans.
5) *Contract termination rights*: In a situation where a force majeure results in prolonged prevention of work performance, the affected party reserves the right to terminate the contract to minimize losses and be free of liability for breach of contract. A lot of instances like this occurred during the prolonged COVID-19 event, and some contracts had to be renegotiated or novated.

Force majeure clauses are essential in construction contracts for addressing unforeseen circumstances that can disrupt the project, its timeline, cost, and quality performance. By defining and specifying the scope of force majeure events and outlining the rights and obligations of the parties in sharing the risks, project teams can better manage consequential risks and facilitate project continuity in challenging events and circumstances. If not well handled, with the contract clauses appropriately interpreted and applied, disputes can arise between the parties in the following ways:

1) *Event classification disputes*: Where a force majeure event occurs, disputes may arise between the parties as to its proper classification among the named events in the contracts. For instance, a Category 1 hurricane may be argued by one party to be a manageable thunderstorm that does not raise the severity of force majeure in the spirit and intent of the contract. Hence, claims under this scenario may be limited, especially where it can be argued that reasonable steps were not taken to mitigate the impacts.
2) *Debatable extent of impact of an event*: The extent of a force majeure event's impact is open to debate, and it can be disputed between the parties. For instance, one party may claim 60 days of time extension, while the other may think the days should be much less, like 20 or 25 days. The same applies to cost overruns and excuses for dropped quality or performance.
3) *Obligation of parties*: The share of risks, liabilities, and obligations can be open to debate and disputations between the parties, as neither is responsible for the event. Hence, the reasonability of claims may be disputed.
4) *Compliance with contractual provisions on force majeure*: Noncompliance with contractual requirements for notification of an event, its extent and impact on performance, and the reasonable mitigation of damages steps taken may result in disputation of claims between the parties over the disruptions and consequences of an event. The burden of proof lies with the party claiming force majeure.

It is also more plausible to avoid disputes from force majeure events rather than to prevent them through having existing cultures and structures in place. Avoiding disputes from force majeure events in construction projects calls for adopting a comprehensive contract with clauses that name and specify all possible events and risks, extents and possible performance impacts, the obligations of the parties, and their sharing of consequential risks, liabilities, and responsibilities, and what amounts to reasonable impact mitigations; collaborative project teams that can promptly deal with the challenges and fallouts from a force majeure event; and maintenance of clear, effective, and open communication to swiftly address and document events and extents, work disruptions, and mitigation efforts applied. Any challenges and event-related issues should be swiftly doused before they escalate to disputes that call for resource-consuming negotiations, mediation, arbitration, or litigation.

Please refer to Table 14.1 for the recommended PDM's D-PAD concept applicable in this regard.

9.1.3 Regulatory Enforcements

Regulatory enforcement refers to the processes through which governmental authorities and designated agencies monitor, enforce, and ensure compliance with standards, laws, regulations, and codes pertaining to construction activities. Governmental authorities' oversight is needed due to several reasons, such as safety standards, structural integrity, environmental protection compliance, quality assurance and performance standards, as well as order. Here is how regulatory enforcements occur in construction:

1) *Enforcement agencies*: They may be governmental agencies or government-designated specialized agencies:
   a) *Governmental agencies*: National, state, provincial, municipal, or local government agencies oversee monitoring, enforcement, and compliance with building standards, laws, regulations, and codes in construction activities. These agencies include the Occupational Safety and Health Administration (OSHA), labor departments, building departments, and environmental protection agencies (EPA).
   b) *Government-designated specialized agencies*: In some jurisdictions, specialized agencies may be designated to oversee certain aspects of construction activities, e.g., fire protection like the National Fire Protection Agency (NFPA), electrical safety such as Public Service Commission, and environmental impact assessments such as Council on Environmental Quality (CEQ).
2) *Areas of regulatory enforcement*: They include:
   a) *Building codes and standards*: The department of building and standards monitors, enforces, and ensures compliance with minimum requirements or codes for construction practices, materials, structural design and integrity, accessibility, fire safety, and energy efficiency.
   b) *Zoning and land use regulations*: Land or plots and locations are designated for certain uses in the overall neighborhood, urban, or regional planning. Authorities oversee compliance with zoning laws, ensuring that only permissible building developments are allowed to be built in their designated areas or zones. For example, at a location zoned residential, it may not be permissible to build a hotel there. The department overseeing zoning also controls the density of developments, setbacks, easements, and environmental impact assessments.

   c) *Health and safety regulations*: OSHA enforces workplace health and safety compliance to protect workers from hazards of accidents, falls, electrical shocks, and safe maintenance and handling of equipment, hazardous substances, and work.
   d) *Environmental regulations*: Regulators oversee the environmental impact of construction activities on people, natural habitats of flora and fauna, soil erosion prevention, pollution of air and water, and siltation of water bodies.
   e) *Licensing, approvals, and permits*: Regulatory authorities issue licenses, approvals, and permits for various aspects and stages of construction activities, such as demolition, site preparation, excavation, building erection, and service installation, such as electrical, plumbing, and HVAC. They issue a Certificate of Occupancy (CO) when the building project is certified as occupiable for its purpose.
3) *Regulatory enforcement methods*: They include:
   a) *Approval reviews and permits*: Authorities review and approve or permit construction activities determined as complaint with zoning ordinances, setback and easement requirements, and other regulatory requirements.
   b) *Inspections*: Authorities or inspectors conduct periodic announced and unannounced visits to construction sites to verify compliance with the required applicable standards and regulations. Inspectors can inspect construction methods relative to required standards, materials, allowances, safety, and environmental compliance.
   c) *Citations, fines, and work stoppages*: Inspectors have the authority to issue citations, levy fines, and order stoppage of work where they observe noncompliance or violations of the standards and regulations relative to their severity.
   d) *Employee and public reporting and complaints*: Employees and the public can report observed noncompliance with regulations and safety standards. The reporting employee cannot be retaliated against in any way by law.

Regulatory enforcement plays a vital part in promoting safe, sustainable, and standards-compliant construction activities. Navigating around the many regulatory requirements for construction activities across jurisdictions and disciplines can be quite complex and challenging for project participants and stakeholders. Regulatory requirements can be evolving, and they can change anytime, requiring ongoing education and adaptation to the changes and the new standards. Regulatory enforcement can be a trigger for disputes due to its potential to impact project timelines and costs. Here are some of the ways regulatory enforcement can trigger disputes:

1) *Licensing, approvals, and permits*: Delays or denials of required licensing, approval, or permits can forestall construction activities, leading to bursting

of project timelines with subsequent cost overruns and disputatious claims. Contractors may dispute the sufficiency of documents provided by owners to obtain the permits, or they may challenge regulatory decisions affecting the progress of the project.

2) *Noncompliance with building codes and standards*: Regulators may issue citations, fines, or work-stop orders due to observed noncompliance with applicable building codes and standards. Disputes can arise over the cause of the violation. Contractors may dispute that they have built in accordance with the documents provided by owners that unfortunately failed to comply with the applicable codes and standards.
3) *Environmental impact assessments and compliance*: Regulators may issue fines, delay, or stop work due to observed environmental protection violations based on environmental impact assessments, pollution control, waste management, and protection of habitat. Disputes can arise between the parties over the cause of the responsibility and liability for noncompliance and the corrective actions.
4) *Occupational health and safety regulations*: OSHA inspectors may issue citations, fines, or work-stop orders due to observed safety violations on construction sites that result in falls, illness, injuries, electrical shocks, exposure to hazardous substances, or death of workers or such fatalities. Disputes can arise between the parties over responsibility and liability for noncompliance with safety regulations, adequacy of implemented safety protocols, and equipment, such as personal protection equipment (PPE), fines, and resultant delays.
5) *Unpredictable evolving regulatory requirements*: When there are changes or amendments to building regulations during a project's construction lifecycle, time-taking and costly work changes, adjustments, and adaptations may be inevitable. Disputes can arise between the parties over the responsibility and liability, or equitability thereof, for the consequential impacts on the project timeline and cost.

It is also more plausible to avoid disputes from fallouts from some regulatory enforcements, even though many are preventable through entrenched internal best practices and cultures. Avoiding disputes from regulatory enforcements in construction projects calls for contractual clarity on the responsibilities for regulatory compliance, obtaining of licenses, approvals, and permits, as well as compliance with the standards; proactive self-checks and quality control (QC) for compliance without waiting for when the authorities come to do so; and maintaining comprehensive and accessible approval and permits documentations, including inspections and corrective actions, if any. Any required corrective actions must be promptly accomplished even before inspectors, if possible, and any challenges and issues should be swiftly doused to avert them from escalating to disputes that call for resource-consuming negotiations, mediation, arbitration, or litigation.

Please refer to Table 14.1 for the recommended PDM's D-PAD concept applicable in this regard.

9.1.4 Politico-economic Changes

Politico-economic changes refer to shifts in political and economic conditions that affect construction activities, some of which are unpredictable. These changes can emanate from government policies, economic trends, global connectedness, transnational and transcontinental epidemics or pandemics, such as COVID-19, that brought vast global economic shifts that affected construction activities, labor, and supply chain. Understanding politico-economic changes is essential for the construction industry, as they impact project feasibility, market dynamics, regulatory environment, and overall business operations. Here are how politico-economic changes can impact construction activities:

1) *Political stability*: Trade policies, international relations, and international supply chains can be affected by political stability, governmental transitions from one leadership to another, and geopolitical tensions. When a new government takes over and changes trade policies, like the introduction or raising of tariffs on imports from competing countries, that can affect the prices of goods and services coming from those countries. Transitions can bring with them changes in governmental priorities and regulatory changes.
2) *Economic trends*: The economy can fluctuate at any time due to both foreseen and unforeseen factors. Such fluctuations may be responded to with interest rate adjustments, some of which can turn a feasible construction project into an infeasible one. Business cycles can change, and inflation can occur from mild to hyper. The economy can grow and shrink with little notice for preparation to absorb the shocks. Consumer confidence can be low in a shrinking economy, and investment in construction projects of certain or all sectors may be negatively affected.
3) *Government policies, priorities, and regulations*: Political changes can bring with them shifts in government policies, priorities, regulations, and legislation directly affecting the construction industry. Public infrastructure spending and appropriations may change, as in zoning laws, building codes, tax policies, and environmental regulations. Design requirements, project approval, permit protocols, construction standards, and operational practices can be affected. New government priorities and policies on sustainable design and construction come with stricter environmental standards in project design, building services, and material choices. More construction projects will take place with increased infrastructural budget and appropriations. The opposite will be the case with decreased infrastructural budget and appropriations.

4) *Adoption of innovation and technological advancements*: Innovations and technological advancements can have significant impacts on construction activities if adopted. Examples are new material development, construction technologies, prefabrication, and digital tools, such as building information modeling (BIM) and clash detection technologies. While adoption of new innovations and technologies can enhance productivity and speed up the construction process, reducing project costs and timelines through increased efficiency, their integration may demand long learning curves and costly training and infrastructure upgrades, raising overhead and affecting project budget and timelines.

Politico-economic changes can blindside project participants and stakeholders and significantly affect the process and outcomes, depending on how they are addressed. If not properly managed, they can be triggers of disputes due to the inherent uncertainties altering project dynamics and impacting the fulfillment of contractual obligations. Disputes from politico-economic changes can come in the following ways:

1) *Public infrastructure budget and appropriations*: There may be funding cuts or project cancellations for public projects already under bid or construction due to governmental transitions and economic downturn, and contractors can dispute reduced payment or nonpayment for claims. Disputed claims may be on bid preparation costs, loss of business, rental costs, and move-out costs.
2) *Economic downturn and market volatility*: Inflation is a regular phenomenon that can quickly increase building material costs, equipment rental, and labor costs. When hyperinflation happens in the middle of a project construction lifecycle, contractors may resort to corner-cutting if there are inadequate contingencies or the owner is unreasonable to share increased costs, and this can lead to disputes between the parties. Contract renegotiations may be disputed, as with cost escalation claims, delays in project financing and payment, or parties' insolvency and payment defaults. Force majeure claims may be disputed if raised in a recession or an inflationary jump.
3) *Supply chain disruptions from political and geopolitical circumstances*: War may break out at locations from where certain materials are sourced, or routes through which those materials pass. An example is the war in the Middle East that affected the passage of goods and services through the Strait of Hormuz, through which many ships laden with building materials and other commodities navigate from Asia to other continents. The COVID-19 pandemic disrupted the supply chain in almost all sectors. The Russia–Ukraine war disrupts the supply chain of various commodities. Disruption of supply chains causes longer lead times, which can stretch project timelines, leading to disputes over liability for consequential damages between the parties. Force majeure claims may be disputed if raised in this situation.

It is also more plausible to avoid disputes from most politico-economic changes, but some are preventable through entrenched internal best practices and cultures of risk assessments and management. Avoiding disputes from politico-economic changes in construction projects calls for contractual clarity on the responsibilities for foreseen and unforeseen politico-economic changes with executable force majeure clauses; allowing adequate contingencies for potential politico-economic changes; providing for flexibility in project planning and scheduling to accommodate potential politico-economic changes; and maintaining clear, effective, and open communication to swiftly address and document politico-economic changes and extents, work disruptions, and mitigation efforts applied. Any challenges and issues should be swiftly doused to avert them from escalating to disputes that call for resource-consuming negotiations, mediation, arbitration, or litigation.

Please refer to Table 14.1 for the recommended PDM's D-PAD concept applicable in this regard.

9.1.5 Differential Lead Times

Lead time in construction refers to the duration it takes to deliver materials, components, and services for a project during procurement and construction from the time an order is placed, or a request is made, to the time when they are available or delivered. Some materials, components, and services have short lead times, while others may have long lead times. It may also be that some are in short supply and will take time to manufacture or deliver. These materials, components, and services differential lead times need to be coordinated relative to the project timeline. Differential lead times can also be dictated by uncertainties from many causes or sources. Hereunder are explanations of how differential lead times can occur in construction:

1) *Materials and equipment*: Different materials and specialized equipment can have varying lead times due to factors that include availability, manufacturing time, shipping logistics, and seasonal demand variations. For instance, ordering steel beams may have longer lead times when compared with procuring concrete masonry units.
2) *Subcontractor services*: Lead times can differ for subcontractor services such as HVAC, plumbing, electrical, carpentry, and other specialty trades. Again, the reason may be due to the availability of skilled hands, commitment to other projects, and scheduling constraints, as well as seasonality in their business cycle.
3) *Design and engineering services*: Lead times can vary for design and engineering services in terms of drawing and specification preparation relative to project complexity, commitment to other projects, and coordination with other project

stakeholders. For instance, some requests for information (RFI) may call for drawings and specifications with other technical documentation, and these will take time to prepare and may need coordination with other project stakeholders.

4) *Approvals and permits*: Lead times can vary for obtaining approvals and permits, and this affects construction activities and the project timeline.

Differential lead times in construction can be triggers of disputes due to their impact on project schedules, costs, and fulfillment of contractual obligations. Disputes from differential lead times can come in the following ways:

1) *Schedule delays*: Materials, components, equipment, or services with long lead times can disrupt and delay project schedules, with consequential damages that can trigger disputes between the parties over who assumes the responsibilities and liabilities. The validity of claims for time extension can be disputed, while the adequacy of procurement and contingency planning can be questioned.
2) *Cost overruns*: Expedited shipping can cost more if adopted to reduce long lead times. More storage spaces may be acquired or rented to mitigate delays in procurement. Overtime hours may be needed to minimize long service lead times. The additional costs for these measures may be disputed by either party.
3) *Sequencing of construction activities*: Subcontractors' availability can disrupt the sequencing of planned construction activities and change the project environment dynamics, breeding an atmosphere for potential conflicts and disputes. Subcontractor delays can lead to cascading delays across the project.
4) *Change of work and scope*: Accommodating changes in lead times and availability of materials, equipment, and services can call for alternative avenues that could mean changes in aspects of the work and scope, as well as new design details and specifications. Revaluating or renegotiating the changed work and scope can be contentious and disputatious between the parties.
5) *Meeting contractual obligations for claims*: Delays in procurement and delivery can skew the project in many ways and may call for a change of aspects of the work and scope to accommodate or navigate around the long lead times for availability. Performing work out of sync with contractually outlined schedules and milestones with performance requirements can be disputed as a breach of contract for not meeting contractual obligations. Claims under this scenario can be disputed by the parties.

Some potential disputes from differential lead times in construction can be avoided, while some are preventable through entrenched internal best practices and cultures of risk assessments, management, and working with robust, proactive procurement and buyout planning with adequate contingencies. Avoiding disputes from differential lead times in construction projects calls for contractual clarity on the responsibilities for predictable and unpredictable differential lead times with clear lead time requirements, delivery schedules, and consequences for delays; providing for flexibility in project planning and scheduling to accommodate changes in lead times; implementing integrated project delivery (IPD) supported with BIM and collaborative planning to streamline coordination, communication, and decision-making across project participants and stakeholders; and maintaining clear, effective, and open communication to swiftly address and document differential lead times and their extent, work disruptions, and mitigation measures or alternatives applied. Any challenges and issues should be swiftly doused to avert them from escalating to disputes that call for resource-consuming negotiations, mediation, arbitration, or litigation.

Please refer to Table 14.1 for the recommended PDM's D-PAD concept applicable in this regard.

9.1.6 Labor Unions

Labor unions play a crucial role in representing the interests of construction workers, negotiating master labor agreements, and advocating for fair wages for work, benefits, entitlements, and work conditions. The second Industrial Revolution that saw a phase of rapid scientific discoveries, standardization, mass production, and industrialization from the late nineteenth century into the early twentieth century, 1870–1914, also witnessed the beginning of labor unionization to better facilitate communication between workers and their employers and obtain better work conditions and considerations. Labor unionism strengthened around World War I and into World War II.

Labor unions can operate at various levels, from local, regional, national, to international. That is, union chapters represent workers of a craft or trade within specific geographic areas to regional councils and national or international organizations that coordinate industrywide initiatives. Some contractors use union workers, and others use both unionized and/or nonunionized workers.

Unions are of two broad types, namely, trade or craft and industrial unions. Trade unions represent specific trades, such as carpenters' unions, electrical unions, or plumbing unions, while industrial unions represent all workers, irrespective of specialty, within a specific industry, such as auto unions, scriptwriters' unions, and hospitality workers' unions. Membership in unions provides access to union benefits, such as representation in labor negotiations, unfair labor practices (ULP) that affect the members, and participation in union activities.

Unions operate democratically and elect officers and representatives who oversee union operations, organize for negotiation with employers, enforce labor agreements, and advocate for member interests. Here is an overview of labor unionism in the construction industry:

1) *Collective bargaining for workers*: Rather than allowing individual workers to negotiate work conditions with their employers, workers are collectively

represented by a union or unions that collectively negotiate work conditions with employers on behalf of workers under what is called a collective bargaining agreement (CBA) or enterprise bargaining agreement (EBA) in some countries. The CBA outlines union member wages, benefits (such as healthcare, retirement, paid and unpaid leaves), work time and hours, safety protocols, and other work conditions or conditions of employment. The unions advocate and fight for workers' rights, ensuring fair labor practices, while interrogating or litigating ULP against employers on behalf of union member workers.

2) *Skilled labor*: In justifying the importance of unions in catering to their members and their work success, many unions operate apprenticeship programs to train and upskill their members in the crafts and trades they cover. Skill levels and standards are established by certifications, ensuring that workers meet required industry skill standards.
3) *Workplace conditions and safety*: Unions advocate for safe and better workplace conditions free of discrimination for all employees. Workers must be provided with the necessary PPE to ensure their safety while at work. Better workplace conditions include a clean and safe work environment, access to restrooms, first aid, and break times.
4) *Job security and benefits*: Unions introduce safety nets for workers and protect them from arbitrary layoffs, which would otherwise be the case for workers without unions. Procedures for handling downsizing are collectively negotiated and agreed between unions and employers on behalf of union member workers. Unions negotiate comprehensive benefit packages, such as healthcare coverages, pension or retirement plans, workers' compensation and disability insurance, as well as other financial protections for workers and their families.
5) *Political advocacy for workers*: Unions lobby the legislator for favorable labor laws and conditions on behalf of their members. Some donate to and endorse political parties that are favorable to unionism. Unions wield political influence on the legislators for fair wages, equal pay, workplace safety, and funding of construction and infrastructure. Unions also engage the community for the benefit of workers and their families.

Labor unions play a critical role in shaping the construction industry standards, promoting skill acquisitions and skilled labor, and ensuring that construction projects are conducted in a safe and conducive environment with respect for workers' contributions and welfare. Nonetheless, labor unions are occasionally the source of disputes in the construction industry due to various reasons related to labor relations, contract negotiations, and adversarial workplace dynamics. Here are some ways in which disputes can be triggered by labor unions:

1) *Contentious or disputatious collective bargaining*: CBAs that involve wages, benefits, and conditions negotiations and renegotiations can often be contentious

or disputatious. Unions and employers may not be negotiating in good faith, and either may apply stalling and hand-twisting tactics. Unions are notorious for using legal union weapons like picketing, strikes, and boycotts to extract better negotiations from the employers. These weapons are certainly adversarial, and they can disrupt work and stretch project timelines with damaging consequences. Such a scenario can lead to disputes. Disputes can arise over stalled negotiations and disagreements over wage increases and benefit contributions, terms of employment, work time and hours, or safety protocols. Failure to timely reach agreements that call for union strikes, boycotts, and picketing, as well as employer lockout of workers, will all impact project timeline and costs, which are triggers of disputes.

2) *Work stoppages from strikes, boycotts, and picketing*: Strikes, boycotts, and picketing disrupt construction projects, leading to delays, increased costs, and contractual penalties. A workers' lockout by employers and engaging other hands on the project can affect workmanship and project quality. Disputes may arise over which party is to absorb the responsibilities and liabilities from union actions and their consequential impacts on the project timeline, cost, and quality.
3) *Disputes from grievance handling between workers and employers*: If grievances between workers and employers cannot be addressed through informal processes, intra- and interparty disputes can arise up to the level of formal negotiation, mediation, arbitration, or litigation. In contention may be disputes over disciplinary actions, work assignments or reassignments, preferential or seniority rights, or compliance with safety protocols.
4) *Differences in interpretation and application of employment and jobsite rules*: Unions may insist on certain parts or all of the job to be handled by union member workers. They may reject some prefabrications. They may specify the number of workers to engage in completing certain tasks. They may demand onerous safety protocols and other labor practices to protect workers. Some of these advocacies may be disputed by the employer as ULP, and this scenario can change the dynamics of the project environment to adversarial, leading to investment of time and efforts to resolve, which can impact project timeline and costs—a major trigger of disputes between the parties.
5) *Impact on effective project management*: Union rules and work practices can significantly affect project management efforts, affecting staffing, scheduling, and productivity. Disputes can be triggered between the parties if union rules and work practices conflict with the project's efficiency and productivity goals, project timelines, costs, and the meeting of contractual obligations. The severity of the impact of union rules and work practices on project timelines and costs can be disputed between the parties.

For contractors using union workers, some potential disputes from labor unions can be avoided, while some are preventable through entrenched internal best practices and cultures of deep and extensive feasibility studies, risk assessments, and management, and working with robust, proactive labor agreement awareness and labor relations. Avoiding disputes triggered by labor unions in construction projects calls for contractual clarity on the responsibilities for predictable and unpredictable labor actions and how to handle them; providing for flexibility in project planning and scheduling to accommodate labor action issues; fostering collaborative problem-solving approach between unions and contractors, as well as project owners to find mutual common grounds, explore creative solutions, and mitigate or avert the potentials of conflicts before they impede project progress; and promoting clear, effective, and open communication to swiftly address and document labor union issues and extents, work disruptions, and mitigation measures. Any challenges and issues should be swiftly doused to avert them from escalating to disputes that call for resource-consuming negotiations, mediation, arbitration, or litigation.

Please refer to Table 14.1 for the recommended PDM's D-PAD concept applicable in this regard.

9.1.7 Labor Seasonality

Labor seasonality in construction refers to the fluctuations in the availability of, and demand for, construction workers throughout the year. Their availability and demand are controlled by several factors like seasonal weather conditions and regional differences. Understanding labor seasonality is critical in effective planning for staffing, managing resources, and mitigating challenges associated with seasonal fluctuations of labor. Here are contributors to labor's seasonal fluctuations in more detail:

1) *Weather conditions*: For regions with distinct seasons, like winter, spring, summer, and autumn (or fall), or dry season, harmattan, and rainy season, weather conditions play a significant role in the extent of construction activities. For instance, harsh winter conditions can limit most outdoor construction work, leading to reduced demand for labor during this period. Some construction workers indulge in hunting during the fall and part of winter, and such indulgence determines their availability for construction work. In contrast, spring and summer tend to witness increased construction activities, and demand for construction workers is higher.
2) *Regional differences*: Regions with milder winters are likely to witness more consistent construction activities throughout the year as compared to regions with harsher winters.
3) *Statutory holidays*: Annual holiday periods affect labor seasonality as workers take these days off work. Construction companies need to plan these times

into their project schedules and staff to accommodate the paucity of workers in these periods and better maintain and predict project timelines.

4) *Vacation periods*: There are periods of the year when many workers like to take time off for vacations, depending on the regions where they find themselves. Some in colder regions may want to take time off work to vacation in warmer regions at any end of the earth. Knowing about popular vacation periods in the regions of construction work is essential in planning better for staffing and adjusting for these periods in working out achievable project timelines.
5) *Project cycles*: Construction project cycles often closely follow seasonal cycles, meaning that there may be more projects and greater demand for labor during some seasons as compared to others. Harsh winter conditions are likely to see shorter project cycles and opportunities compared to warmer periods.
6) *Project owner preferences*: Most project owners tend to prefer to engage in construction during more conducive seasons due to reduced challenges and uncertainties, which can affect project schedules, construction methods, and costs.
7) *Economic trends*: The regional economy can also fluctuate with seasons, whereby more economic activities take place during more conducive seasons due to reduced challenges and uncertainties, which can see increased activities in all sectors with multiplier effects that trickle down to increased construction activities and demand for construction workers.

Overall, labor seasonality can present serious challenges to the construction industry, including unhelpful labor shortages, cost fluctuations, uncertainties, resource management headaches, and the need to invest in continuous training and retention of workers. Due to impacts on workforce availability, project timelines, and fulfillment of contractual obligations, labor seasonality can contribute to disputes in construction in the following specific ways:

1) *Labor shortages and increased costs*: Based on the law of supply and demand in economics, all things being equal, when there is a decrease in the supply of construction workers, and demand for them is the same or high, their costs tend to rise to higher equilibria. Contractors compete for the decreased available number of workers and are willing to pay higher wages and costs to beat competitors. Disputes may arise between the parties due to cost escalations or contractors cutting corners and lowering quality to make up for the increased costs. Contractors asking for reimbursements for increased costs may be disputed by project owners.
2) *Project progress interruptions and delays*: Labor scarcity during peak construction seasons due to any reason (seasonal fluctuations or otherwise) would mean that contractors cannot keep to schedule and will experience delays in the completion of project milestones and timelines. Project owners may not understand

the resultant delays or see them as justified, and contractors may find their claims of time extension disputed.

3) *Productivity and quality disputes*: When there is a scarcity of labor, the chances increase of contractors making do with any hands that are available, irrespective of skill levels. Productivity and workmanship quality will suffer under these circumstances, and it can be a trigger of disputes between the parties. Inexperienced or temporary workers can be hired in this scenario to make up for the shortage of skilled labor. Remedial and rework can follow, both of which can be subjects of disputes between the parties.
4) *Meeting contractual obligations*: It is rare to see a construction contract that takes labor seasonality into consideration. Hence, meeting contractual obligations is expected regardless of labor seasonality. Meeting contractual deadlines and milestones is expected. Failing to meet them would amount to a breach of contract that can trigger disputes between the parties. Project owners can invoke liquidated and ascertained damages clauses or contractual penalties for failing to meet contractual deadlines and exceeding the project timeline.
5) *Safety and compliance concerns*: Amid the scarcity of skilled labor, engaging inexperienced or temporary workers to make up the gap can compromise safety protocols and compliance with standards and regulatory requirements. Citations, fines, corrective mandates, or work stoppages that result from any violations can potentially be triggers of disputes between the parties due to their impact on project cost, timeline, and disagreement over which party should take part or all of the responsibilities and liabilities.

Some potential disputes from construction labor seasonality can be avoided, while others are preventable through entrenched internal best practices and cultures of robust feasibility studies, risk assessments and management, and working with proactive planning and forecasting with adequate contingencies. Avoiding disputes from construction labor seasonality calls for contractual clarity on the responsibilities for predictable and unpredictable construction labor seasonality risks, cost escalation and time extension entitlements that may be applicable; provision of flexibility in project planning and scheduling to staffing and labor availability fluctuations; maintaining clear, effective, and open communication to swiftly address and document labor availability fluctuation issues and extents, work disruptions, and mitigation measures. Any challenges and issues should be swiftly doused to avert them from escalating to disputes that call for resource-consuming negotiations, mediation, arbitration, or litigation. By understanding potential disputes that can be triggered by construction labor seasonality and implementing proactive strategies that PDM proposes to manage workforce fluctuation challenges, project participants and stakeholders can mitigate risks from surprises and uncertainties, maintain project progress, and foster the necessary collaborations to achieve successful project outcomes.

Please refer to Table 14.1 for the recommended PDM's D-PAD concept applicable in this regard.

9.2 Conclusion

External factors such as weather conditions, force majeure events, regulatory changes, and economic fluctuations remain among the most unpredictable yet impactful triggers of disputes in construction projects. Their influence often extends beyond technical considerations, affecting timelines, costs, and stakeholder relationships. This chapter underscores that while these factors cannot be controlled, their risks can be anticipated and mitigated through proactive strategies. Applying PDM principles—D-PAD enables project teams to integrate contingency planning, robust communication protocols, and adaptive scheduling into the project lifecycle. By conducting early dispute-risk analysis (DRA) and fostering collaborative responses, stakeholders can transform external uncertainties into manageable challenges rather than costly disruptions. Ultimately, embracing foresight and flexibility in addressing external factors strengthens resilience, safeguards project objectives, and promotes harmony among all participants.

Chapter Review Questions

1) External factors in construction projects are always predictable and manageable. True/False?
2) Weather conditions such as rain, snow, and heat can disrupt construction schedules and safety. True/False?
3) Force majeure events are typically covered in construction contracts to excuse delays. True/False?
4) Regulatory enforcement can lead to disputes if permit processes are delayed or changed. True/False?
5) Labor unions have no influence on construction project timelines or costs. True/False?
6) Economic downturns can affect the financial stability of contractors and suppliers. True/False?
7) Labor seasonality refers to the consistent availability of workers throughout the year. True/False?
8) Which of the following is an example of a force majeure event?
   a) Late payment by the owner
   b) A subcontractor dispute

    c) A hurricane disrupting site access
    d) A change in project scope
9) What is a common result of differential lead times in material procurement?
    a) Faster project delivery
    b) Coordination challenges and delays
    c) Reduced costs
    d) Improved safety
10) Which external factor can cause sudden changes in building codes or permit requirements?
    a) Labor seasonality
    b) Regulatory enforcement
    c) Weather conditions
    d) Material shortages
11) Politico-economic changes can affect construction projects by impacting:
    a) Site layout
    b) Material quality
    c) Financing and scheduling
    d) Team communication
12) Which of the following best describes labor seasonality?
    a) A change in labor laws
    b) A fluctuation in workforce availability due to seasonal demand
    c) A union strike
    d) A change in contract terms
13) What is a recommended strategy to manage disputes arising from external factors?
    a) Ignore external risks
    b) Rely solely on internal controls
    c) Conduct early risk analysis and include flexible contract clauses
    d) Avoid communication with stakeholders
14) Which external factor can lead to disputes over material pricing and availability?
    a) Labor union activity
    b) Politico-economic instability
    c) Regulatory enforcement
    d) Force majeure
15) ________ factors are influences outside the control of the project team that can affect construction outcomes.
16) The legal term ________ refers to events like natural disasters that excuse contractual obligations.
17) ________ conditions such as wind, snow, and heat can delay construction and increase safety risks.

18) ________ enforcement refers to government actions that may alter or delay project approvals.
19) ________ lead times refer to varying durations for material procurement and delivery.
20) ________ seasonality affects labor availability based on seasonal employment trends.

Exercises

Exercise 9.1 Force Majeure Clause Simulation

Objective: Evaluate how force majeure events impact contractual obligations and dispute resolution.

Instructions:
Form small groups and assign roles: Owner, Contractor, Legal Advisor.
Present a scenario: A major flood halts construction for three weeks.

Each group must:
- Review a sample contract's force majeure clause.
- Determine whether the event qualifies under the clause.
- Decide how responsibilities, costs, and schedule adjustments should be handled.
- Draft a short addendum or resolution plan.

Outcome: Understand how external events are interpreted legally and practically in construction contracts.

Exercise 9.2 Regulatory Shock Response Drill

Objective: Practice responding to sudden regulatory enforcement changes during a project.

Instructions:

Scenario: Midway through a hospital construction, new fire safety regulations are introduced, requiring design changes.

In pairs or teams, develop a response plan that includes:
- Communication strategy with stakeholders.
- Budget and schedule impact analysis.
- Dispute avoidance or dousing measures.
- Contractual adjustments or renegotiation steps.

Outcome: Learn to manage regulatory disruptions proactively and collaboratively.

Discussion

1) How do political and economic changes—such as inflation, trade restrictions, or policy shifts—affect construction projects? What strategies can contractors and owners use to protect themselves from these external risks?
2) Labor unions can influence project timelines, costs, and workforce availability. In what ways can union activity lead to disputes, and how should project teams prepare for or respond to labor-related disruptions?

10

Delphi Study of Proactive Dispute Management

Readers' Learning Objectives and Outcomes

1) *Analyze the various flashpoints and triggers of conflicts and disputes in the industry beyond project owners and contractors.*
2) *Conceptualize plausible prevention, avoidance, or dousing approaches that may be applicable to these conflicts' and disputes' flashpoints and triggers.*
3) *Match and apply prevention, avoidance, or dousing strategies effective or suitable to each conflict's and dispute's flashpoints and triggers.*

10.1 Introduction

The Delphi method is a structured, iterative process used to gather and refine expert opinions through a series of questionnaires interspersed with controlled feedback. Originally developed by the RAND Corporation in the 1950s for forecasting military and technological developments, the method has since been widely adopted in fields requiring consensus on complex, uncertain, or emerging issues (Beiderbeck et al. 2021; Gutierrez-Bucheli et al. 2024; Hallowell and Gambatese 2010; Linstone and Turoff 2002; Mollaoglu et al. 2015).

In the context of construction and related industries, where disputes are multifaceted and often influenced by behavioral, contractual, and environmental variables, the Delphi method offers a rigorous approach to distill expert insights. It enables a diverse panel of professionals, ranging from owners, contractors, consultants, legal experts, to academics to anonymously contribute their perspectives, reflect on group feedback, and converge toward a shared understanding of best practices. It is used in this book to develop dispute management best practices and to illustrate case studies to industry practitioners.

10.1.1 Justification for Using Delphi in PDM Research

The application of the Delphi method in this chapter is both timely and strategic. Proactive dispute management (PDM), as a relatively novel framework, benefits from the collective wisdom of seasoned professionals who have navigated the complexities of construction disputes. Unlike traditional empirical methods that rely on historical data, the Delphi approach is forward-looking and ideal for exploring emerging paradigms like dispute prevention, avoidance, and dousing (D-PAD).

Key justifications include:

- *Complexity of disputes:* Construction disputes are rarely caused by a single factor. The Delphi method allows for the exploration of interrelated causes and the identification of nuanced strategies for prevention and mitigation. For instance, Kermanshachi et al. (2020) used the Delphi method to identify and rank thirty complexity indicators in industrial construction projects. They demonstrated how disputes often stem from interrelated factors, such as change orders, technology familiarity, and stakeholder coordination. The iterative Delphi process allowed experts to weigh factors and develop a complexity management model, showing how nuanced dispute causes can be systematically addressed.
- *Need for consensus:* As PDM seeks to shift industry norms away from reactive dispute resolution, consensus among experts is critical for legitimacy and adoption. A comprehensive review of 88 Delphi-based studies in construction engineering and management highlights the method's effectiveness in building

expert consensus on complex, uncertain topics (Ameyaw et al. 2016). The authors emphasize that many CM challenges, such as planning, forecasting, and dispute resolution, require collective expert judgment due to limited empirical data. The Delphi method's structured rounds of feedback allow experts to converge on shared priorities and strategies, making it ideal for establishing industry-wide norms in PDM.

- *Anonymity and iteration:* The method's structure encourages honest input, reduces the influence of dominant personalities, and refines ideas through multiple rounds of feedback. Strong guidance suggests the importance of anonymity and iterative feedback in Delphi studies (Khodyakov et al. 2023). In construction, where hierarchical dynamics can suppress open dialogue, anonymity allows experts to share candid insights. Iteration refines these insights over multiple rounds, reducing bias, and it can enhance the quality of dispute-related recommendations.
- *Scalability and adaptability:* Delphi findings can inform policy, training, and implementation strategies across various project types and organizational contexts. Naji et al. (2024) applied the Delphi method to identify and rank 48 dispute-related risk factors during the preconstruction phase of projects. The study demonstrates how consensus among eighteen diverse construction professionals can be systematically achieved to inform proactive dispute strategies and operationalized into practical tools for dispute mitigation.

By employing the Delphi method, this chapter aims to validate and enrich the PDM framework, identify practical barriers and enablers, and offer actionable insights for industry-wide transformation.

10.2 Methodology and Panel Composition

A Delphi panel brings together a pool of experts to tap into their unique experiences and to converge on shared understandings of topics. The Delphi Study in this book consisted of two survey rounds. The questions for round one of the survey gathered demographic information, participants' experience with dispute management, and dispute case studies. As part of the dispute case studies, participants were asked to provide four dispute cases (i.e., most recent, most common, highest escalated, and most expensive) and to assign the dispute to a specific category (e.g., owner-related dispute flashpoint, contractor-related dispute flashpoint, or miscellaneous-related dispute flashpoint) and subcategory (e.g., addenda, labor productivity fluctuations, or quality of drawing documents). They also provided information relating to the type of project delivery method used on the project, summarized the problem, gave the actual cause(s) of the dispute, reported how the dispute was resolved, and the approximate cost to resolve the dispute. Then, the participants mapped their four dispute cases to PDM's D-PAD framework by indicating and noting how each dispute might have been prevented, avoided, or doused.

For round two of the Delphi study, the study participants responded to follow-up questions aimed to reach consensus as to the most effective method of preventing, avoiding, or dousing disputes that might occur due to failures or lapses in the key construction aspects or subcategories discussed earlier in PDM's framework.

Participants in the Delphi study were recruited from a general contractor/construction management company in the Midwest region of the United States. The participation criteria were: (1) work as a project manager (PM) or above (e.g., superintendent, senior PM, director, or executive), (2) have leadership experience, and (3) navigated conflicts or disputes during one's career. An invitation to participate was sent to 30 potential study participants (i.e., 6 from each work group). The number of respondents in round 1 was 13, and in round 2 there was a total of 26 respondents.

The study and its data collection methods were approved by the Institutional Research Board at Purdue University (IRB#2024-1121).

10.2.1 Key Themes Explored

The Delphi study was positioned around our D-PAD framework. As such, the themes derived from the analyses are centered on the following categories:

1) *Owner-related disputes flashpoint*
   a) Unclear/incomplete addenda
   b) Differing site conditions
   c) Change order or variation order by owner
   d) Scope creep
   e) Project turnover or handover problems
   f) Construction acceleration, lofty and unrealistic expectations
   g) Payment delays
2) *Contractor-related dispute flashpoints*
   a) Liquidated and ascertained damages
   b) Progress or interim payments and their possible misuse
   c) Concerns with character and/or reputation of contractor
   d) Lack of technical capability and/or experience
   e) Underbidding and accuracy of estimates
   f) Schedule/productivity concerns
   g) Lack of project resourcing and balancing
   h) Use of union and nonunion labor

   i) Work quality
   j) Commissioning issues
   k) Accidents and safety
   l) Mechanics lien
3) *Miscellaneous dispute flashpoints*
   a) Conflicting work schedules and coordination (subcontractors/vendors)
   b) Discrepancies in contract documents
   c) Sharing of risks and rewards
   d) Litigious nature of participants
   e) Communication and miscommunication
   f) Nature and uniqueness of projects
   g) Rate of requests for information (RFIs)
   h) Poor detailing of novel structures
   i) Unforeseen conditions or circumstances
   j) Weather conditions
   k) Team dynamics/project culture

10.3 Findings and Case Studies

10.3.1 Delphi Study Round One

A kickoff meeting was held for round one of the study with each of the respective groups to introduce the research goals and objectives and to review the survey. In all, there were 13 complete responses to the survey out of 30 total participants (Figure 10.1). The majority of the PMs completed round one of the study.

The study also aimed to have a good distribution across individuals' experience levels. Based on the data, many of the participants had over 20 years of experience

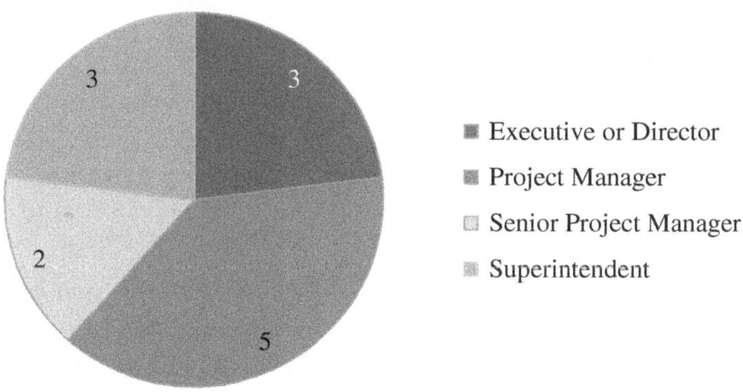

Figure 10.1 Participants' role or position in the company.

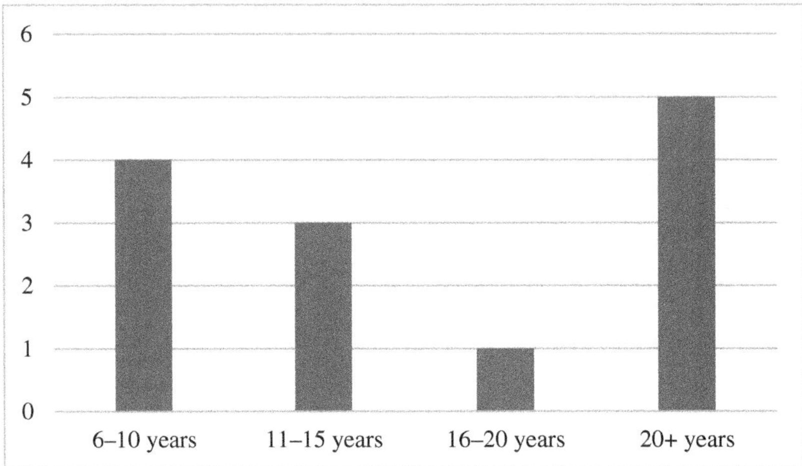

Figure 10.2 Number of years participants have worked in construction.

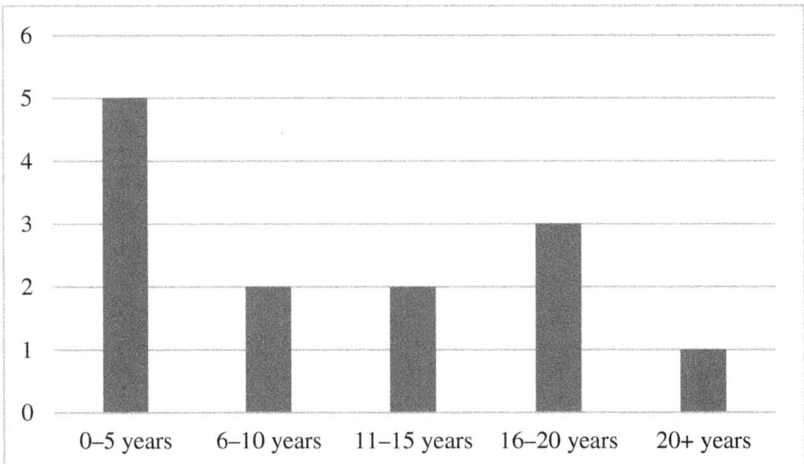

Figure 10.3 Number of years participants worked as a supervisor, project manager, or above.

working in construction (Figure 10.2). Meanwhile, 30% of the respondents had at least 11 years of experience, and another 30% had at least 6 years of experience. Hence, the data captures a rich perspective of seasoned construction professionals with vast experience. The bulk of the participants surveyed have worked in a leadership role for six years or more (Figure 10.3).

Table 10.1 Number of contested disputes anticipated in a project (start to finish).

Contested disputes on a project	# of responses	Average # of disputes
Dispute resolved during the normal course of business	13	35
Dispute resolved by project team meeting	13	21
Dispute resolved by CEO meeting and/or negotiation with associated stakeholders (owner, subcontractor, etc.)	13	4
Dispute resolved by mediation	3	<1
Dispute resolved by arbitration	6	<1
Dispute resolved by litigation	3	<1

As the study homes in on construction disputes, it becomes clear that most disputes are managed at the project level by the project team (Table 10.1). According to the data, a project team might encounter as many as 35 disputes regularly during a project, with some of these being addressed in team meetings. Only a few of these disputes rise to a critical level that requires them to involve the CEO or executive leadership in the negotiation or resolution process. Furthermore, the odds of a dispute moving into full mediation, arbitration, or litigation are even less likely according to the findings. This also considers that the average number of projects the study participants worked on over the last 5 and 10 years ranges from 7 to 15, respectively. Therefore, it can be surmised that most project durations are approximately 18–19 months.

Next, the individual case studies from the study participants are reported. For each of the different projects that had disputes, the participants were to indicate the dispute category and the contract type used, summarize the problem(s), state the actual cause(s) of the disputes, indicate how the dispute(s) were/was resolved, how long each dispute took to resolve, and the costs in dollars (if known) to resolve the dispute(s).

10.3.2 Case Studies

Case 10.1

(Most Recent Example)

Dispute Category: Construction documents-related dispute flashpoint—reading and interpretation of construction documents.

(Continued)

Case 10.1 (Continued)

Project Delivery Method: CM at-risk.

Problem: The fire alarm system installed on the project did not meet the code or the client's requirements.

Actual Cause(s) of the Dispute: The fire alarm system was specified as a delegated design system. In a delegated design, the design professional establishes the overall performance requirements for the system, which must be finalized by a licensed fire alarm contractor. The fire alarm contractor is responsible for the final design, including placing the devices, specifying wiring requirements, and submitting it for approvals. In this case, the fire alarm contractor did not explicitly follow the construction documents, adding additional audible alarms to meet the local code and the client's requirements. The client's insurance company correctly approved the submitted shop drawings, but the architect/engineer (A/E) of record marked up the shop drawings to match the number of audible alarms as shown on the original construction documents. The subcontractor revised their shop drawings to match the A/E's comments but never resubmitted them to the insurance company for approval. The contractor installed the fire alarm system per the revised shop drawings, subsequently failing their final inspection for occupancy. The fire alarm subcontractor had to go back and install all the deleted devices in a finished facility, which added complexity and time for rework.

How the Dispute was Resolved: A meeting was held with the A/E firm, owner, contractor, and subcontractor. The owner agreed to pay the claim based on their A/E marking up the shop drawings.

Costs ($) to Resolve the Dispute: $83,000

How Could the Dispute Have Been Prevented, Avoided, or Doused?

Prevented: The issue could have been prevented by the fire alarm contractor resubmitting the marked-up shop drawings to the insurance company for second review/approval. If the A/E of record had not modified the delegated design, this dispute would likely have been prevented.

Avoided: The construction manager could have reviewed the marked-up shop drawings and questioned the changes, knowing that it was a delegated design. A meeting could have been held to review the mark-ups with all parties involved.

Doused: The fire alarm subcontractor could have raised awareness when they received the mark-up shop drawings—as a preferred vendor to the client, they should have known the requirements based on similar projects working with this owner.

Case 10.2

(Most Common Example)

Dispute Category: Construction documents related dispute flashpoint—quality of drawing documents

Project Delivery Method: CM

Problem: Contractor completed a $60 million construction project with extensive high-end finishes throughout the building. The details for many of the pieces of art and high-end finishes were not constructable because the fasteners did not work as designed. Different fasteners and trim pieces were needed to complete the installation, which were not shown on the construction documents. All the while, there were multiple remobilizations by as many as seven different trades for this scope of work.

Actual Cause(s) of the Dispute: The finish details provided in the construction documents were neither feasible nor adequate for this installation.

How the Dispute was Resolved: The contractor and subcontractors created alternative details that were submitted, reviewed, and approved by the A/E after several meetings, RFI's, and submittal revisions.

Costs ($) to Resolve the Dispute: $85,000

How Could the Dispute Have Been Prevented, Avoided, or Doused?

Avoided: This dispute could have been avoided if the designers had been involved in the building information modeling (BIM) process and workshops to discuss the installation details. Many conflicts associated with this type of dispute can be avoided with more involvement from the A/E team.

Case 10.3

(Highest Escalated Example)

Dispute Category: Owner-related dispute flashpoint—scope creep

Project Delivery Method: Design-bid-build

Problem: The general contractor budgeted the construction costs based on the design documents (approximately 10% complete) and established a contract value with the owner based on similar project experience and expectations.

Actual Cause(s) of the Dispute: The construction documents that were issued included much more scope than what was originally budgeted. The owner added a lot more items to the project, which drove up the price above and beyond the agreed-upon contract.

(Continued)

Case 10.3 (Continued)

How the Dispute was Resolved: The CEOs from each organization (i.e., owner and contractor) held a meeting to resolve differences.

Costs ($) to Resolve the Dispute: $1 million

How Could the Dispute Have Been Prevented, Avoided, or Doused?

Prevented: This could have been prevented by establishing protocols and practices to not enter into a contract agreement based on incomplete drawings. This became clear early on, once the final construction documents were issued, showing a variance between the scope and costs.

Avoided: This could have been avoided by waiting for the final construction documents to be issued before providing any budget pricing, despite pressures from the owner to offer pricing sooner to get the project started.

Doused: This dispute can be doused by bringing the parties together immediately to resolve the discrepancy, resulting in a change order to capture the costs of additional scope.

Case 10.4

(Most Expensive Example)

Dispute Category: Owner-related dispute flashpoint—scope creep

Project Delivery Method: CM at-risk

Problem: The contractor entered a contract to build a $60 million high-rise hotel construction project. By the end of the project, the construction value was $90 million due to the added scope of work that was tracked through construction change directives (CCDs), architects' supplemental instructions (ASIs), bulletins, and RFI's. The owner pressured the contractor to continue with the work without approving change orders to maintain the construction schedule.

Actual Cause(s) of the Dispute: The construction documents were poorly detailed, which led to many assumptions in the original estimates.

How the Dispute was Resolved: Despite multiple meetings with the CEOs from the respective organizations (i.e., owner and contractor), the disagreement ended up in arbitration.

Costs ($) to Resolve the Dispute: $25 million

How Could the Dispute Have Been Prevented, Avoided, or Doused?

Avoided: This could have been avoided if the contractor held regular meetings with the owner to review/approve changes before the work was completed, regardless of the schedule impacts.

10.4 Synthesis of Expert Consensus from Round Two

The findings in this section are derived from 26 experts who completed round 2 of the Delphi study. The experts were distributed nearly evenly among the groups, i.e., six superintendents, six PMs, seven senior project managers, and seven executives or directors. As previously mentioned, the participants were asked to determine the most effective method of *preventing*, *avoiding*, or *defusing* disputes that might occur because of failures or lapses in key construction aspects. Their responses were constrained into three categories: (1) Owner-related disputes, (2) contractor-related disputes, and (3) miscellaneous disputes.

10.4.1 Owner-Related Dispute Flashpoints

These types of disputes invariably show up during the preconstruction or construction stages of a project. They are largely associated with construction project owners, yet they must be managed and understood by all parties involved in a construction project, such as the architects, engineers, general contractors (GCs), and subcontractors.

According to the study participants, the majority of the disputes attributed to owner decisions are preventable except for differing site conditions (Table 10.2). These elements of CM can often be prevented when project teams establish open communication and clarity early on with the owner and owner representatives. The most distinct area where conflicts and owner disputes can be prevented is with addenda. Project teams should follow a robust review process that includes quality assurance (QA) or quality control (QC) procedures for managing all documents and addenda, along with a plan to review addenda with everyone, including the owner, before they are fully released. Conversely, issues related to differing site conditions commonly have to be managed after the fact. In other words, these

Table 10.2 Owner-related dispute categories and best D-PAD approach.

	Prevent	Avoid	Douse
Unclear or incomplete addenda	✓ 81%	✗ 8%	✗ 12%
Scope creep	✓ 65%	! 38%	✗ 23%
Project turnover or handover problems	✓ 62%	✗ 38%	✗ 38%
Change order or variation order by owner	✓ 54%	✗ 23%	✗ 27%
Construction acceleration, lofty and unrealistic expectations	✓ 50%	✗ 23%	! 35%
Payment delays	✓ 46%	✓ 50%	✗ 35%
Differing site conditions	! 38%	✗ 27%	✓ 46%

are doused because earlier strategies to mitigate against unforeseen conditions have failed. Meanwhile, payment delays tend to be something that can be largely avoided or prevented with established and agreed-upon payment processes, clear and consistent communication, transparency, and a plan to address any payment delays.

10.4.2 Contractor-Related Dispute Flashpoints

Disputes and conflicts occur not only between GCs and owners but also among GCs and subcontractors. These challenges can stem from issues related to any of the top four construction metrics: costs, schedule, quality, or safety. Furthermore, they can happen at any time during the construction cycle. Despite this, most of them, according to the Delphi study participants, are preventable (see Table 10.3).

After inspecting the data, it becomes evident that problems attributed to areas such as liquidated or ascertained damages are preventable. As an example, any liquidated damage (LD) or ascertained damage clauses should be highlighted and discussed during bidding and contract negotiation to ensure everyone is on the same page. The GC and subcontractors should also agree upon a realistic baseline schedule that accounts for some level of contingency. And when delays do arise,

Table 10.3 Contractor-related dispute categories and best D-PAD approach.

	Prevent	Avoid	Douse
Liquidated and ascertained damages	✓ 77%	✗ 27%	✗ 12%
Work quality (omissions, defects, failures, rework, punch list, etc.)	✓ 58%	! 42%	✗ 23%
Accidents and safety (EMR, DART, etc.)	✓ 58%	✗ 35%	✗ 42%
Commissioning issues	✓ 58%	✗ 23%	! 35%
Underbidding and accuracy of estimates	✓ 54%	! 38%	✗ 19%
Progress or interim payments and their possible misuse	✓ 50%	! 31%	✗ 15%
Mechanics lien	✓ 46%	! 35%	✗ 15%
Concerns with character and/or reputation of contractor	! 42%	✓ 50%	✗ 35%
Lack of project resourcing and balancing (labor, materials, and equipment)	! 38%	✗ 27%	✓ 50%
Schedule or productivity concerns	✓ 42%	! 38%	✗ 31%
Lack of technical capability or experience	✗ 35%	! 38%	✓ 42%
Use of union and nonunion labor	✓ 42%	✓ 35%	✗ 12%

they must be tracked, documented, and communicated to all parties immediately. The notion is to be accountable when delays do occur and to work collaboratively to provide course corrections.

Three other areas surfaced during round 2 of the Delphi survey (see Table 10.3), where prevention is the most optimal solution. The first is with accidents and safety, where 58% of the participants believe that conflicts and disputes are best managed through prevention by prioritizing safety, fostering a safety culture, monitoring and enforcing safety practices with consistency, and promptly responding when incidents arise while being transparent about the root causes and offering lessons learned to eliminate future occurrences.

Like safety, both commissioning (Cx) and work quality had 58% of the respondents report that dispute flashpoints are commonly preventable when appropriately addressed early on. When commissioning is included as part of the contractual obligations, the project participants should start planning and setting expectations during the design and submittal review phases of a construction project. At this point, the commissioning agent (CxA) for the project should develop a detailed commissioning plan that not only outlines responsibilities and expectations for the parties involved in the Cx process (e.g., CxA, owner, A/E team, contractor, subcontractors, vendors) but also one that they agree upon. The keys to success and avoiding disputes during Cx are communication, being proactive and responsive throughout the process, and aligning Cx with project closeout to ensure that the highest quality standards are met before owner turnover. And while there are a myriad of procedures and practices involved in Cx, for brevity, the findings highlight that communication is paramount.

As for work quality, it goes hand in hand with Cx processes and practices. There is no substitute for a robust QA/QC process, even when commissioning is not part of the contract. According to the respondents, expectations should be communicated early through practices such as preinstallation meetings, mockups, and work inspections when new items are first installed, so benchmarks can be established for subsequent work. This should be an ongoing process using continuous monitoring and quality checks, documentation, feedback, and corrective actions, aiming to prevent rework. But it can lead to disputes because quality can be subjective without clear and measurable metrics or standards. Hence, this is why these types of dispute flashpoints are most preventable, but they require preplanning and formalized QC programs.

10.4.3 Miscellaneous Dispute Flashpoints

There are other areas where dispute flashpoints may arise during the lifecycle of a construction project. They might be attributed to subcontractor or supplier issues or related to the construction documents, the contract terms and conditions, team

Table 10.4 Miscellaneous dispute categories and best D-PAD approach

	Prevent	Avoid	Douse
Unforseen circumstances	✗ 8%	✗ 19%	✓ 73%
Force majeure	✗ 19%	✗19%	✓ 62%
Descrepancies in contract documents	✓ 54%	! 38%	✗ 23%
Team dynamics or project culture	✓ 50%	✗ 38%	✗ 38%
Nature and uniqueness of project (location, site conditions, etc.)	✓ 50%	✗ 19%	✗ 19%
Weather conditions	✗ 23%	! 35%	✓ 46%
Conflicting work schedules and coordination (sub or vendor)	✓ 46%	✗ 31%	✗ 27%
Poor detailing of novel structures	✓ 46%	✗ 27%	✗ 19%
Sharing of risks and rewards	✓ 42%	✗ 27%	19%
Communication and miscommunication	✓ 38%	✓ 38%	✗ 27%
Litiguous nature of participants	✓ 38%	✗ 31%	✗ 31%
Rate or request for information (RFIs)	✓ 38%	! 31%	✗ 27%

dynamics, or even the nature and uniqueness of the project. Many of the disputes encountered in these areas can be mitigated or avoided using strategies to prevent them from happening in the first place. Preplanning and proactive coordination are essential tools that can prevent disputes among project teams. However, some are simply outside the control of anyone on the project team, such as those conflicts or disputes that arise due to unforeseen site conditions.

According to the findings, both unforeseen conditions and circumstances must be doused when dispute flashpoints flare up (Table 10.4). In these situations, all the relevant project participants and stakeholders must respond expeditiously with a spirit of collaboration and transparency to avoid blaming or finger-pointing. Unforeseen conditions are far from predictable, yet contingencies (e.g., costs and mitigation plans) can be put in place should they arise. And when they do, the team should decisively resolve the problem or issue before it erodes the team spirit—that means working collaboratively, adjusting and/or monitoring schedule impacts, and developing a clear resolution plan including cost implications. It is also important to capture lessons learned from these occurrences to benefit future projects.

Similarly, disputes are unpredictable and come about with events such as natural disasters and pandemics. This was the second-highest rated dispute category where dousing applies. These types of events require a balance of contractual foresight, rapid response, and, again, collaborative problem-solving. This was made evident during the COVID-19 pandemic that took the world by storm—wreaking havoc on business continuity, supply chains, and the workforce. While rare, these situations demand that project teams be informed, adaptable, and aligned in their approaches to keep people safe, in addition to completing their contractual obligations. The reality is that no one wins during these situations, so the goal is to focus on fair and equitable solutions that keep the project moving forward, ultimately, to the satisfaction of the owner.

10.5 PDM Implications for Practice

PDM, as a philosophy and practice, is pivotal to navigating conflicts and disputes in the construction industry. Through the Delphi study, we compiled a thematic list of best practices for practitioners:

1) *Best Practices for Owner-related Dispute Flashpoints*
    a) *Addenda Management*
        i) Upfront clarity and direction
            - Provide clear, early instructions from the owner to the design team to reduce the need for addenda.
            - Ensure the initial design documents are complete and reflect owner input.
        ii) Proactive and transparent communication
            - Engage in open, blame-free discussions with the design team and owner to identify and resolve gaps.
            - Use preconstruction (Precon) and owner, architect, and contractor (OAC) meetings to align on scope and address missing information.
        iii) Rigorous review process
            - Conduct constructability and biddability reviews before issuing documents.
            - Review draft addenda with the design team and owner to ensure accuracy and completeness.
            - Use QA/QC processes to catch and correct errors before and after issuance.
        iv) Strategic use of addenda
            - Treat addenda as a tool for clarification, not a fix for incomplete design.
            - Use preliminary versions (e.g., "bootleg" copies) for internal feedback before formal release.
            - If necessary, issue follow-up addenda to correct or clarify previous ones.

v) Documentation and tracking
- Maintain a running log of issues and resolutions visible to all stakeholders.
- Keep detailed records of clarifications and decisions to support transparency and accountability.

b) *Managing Differing Site Conditions*

i) Prevention through preparation
- Conduct thorough preconstruction investigations (e.g., soil borings, GPR, utility locating).
- Include site visits, existing condition reviews, and geotechnical reports in early planning.
- Educate owners on the value of upfront investigations, even if they increase initial costs.

ii) Plan for the unexpected
- Build in allowances, contingencies, and unit pricing to handle unforeseen conditions.
- Include investigative scopes in bid packages when full preconstruction data isn't available.

iii) Immediate and informed response
- Act quickly when conditions differ, especially if they impact the critical path.
- Gather input from subcontractors and field teams to understand the issue and propose solutions.
- Present objective, factual data to owners and designers to support collaborative decision-making.

iv) Communication and documentation
- Maintain open, regular communication with all stakeholders.
- Keep detailed records of site conditions, changes, and discussions to avoid disputes.
- Use OAC meetings and field reviews to keep owners informed and engaged.

v) Collaborative problem-solving
- Approach differing conditions as a team challenge, not a conflict.
- Come to discussions with proposed solutions and cost/time impacts to streamline decisions.
- Focus on mitigation and resolution, not blame.

c) *Managing Owner-Initiated Change Orders*

i) *Early and clear communication*
- Notify the owner as soon as a potential change arises.
- Use regular check-ins, page turns, and pricing workshops to avoid surprises.

- o Ensure the owner understands the impacts of changes on cost and schedule.

ii) *Provide context and clarity*

- o Present clear, factual, and contextual information about the change.
- o Explain the "why" behind the change, including benefits, drawbacks, and implications.
- o Tailor communication to the owner's level of construction knowledge.

iii) *Structured change management process*

- o Establish a defined process for change orders, including required documentation and approvals.
- o Use accurate logs, scopes, and contracts to track and validate changes.
- o Vet all pricing and scope adjustments before presenting them to the owner.

iv) *Collaborative and proactive approach*

- o Engage the owner, design team, and trades early to align on expectations.
- o Understand the owner's intent and work quickly to translate it into actionable documents.
- o Offer multiple cost-effective solutions when possible.

v) *Accuracy and professionalism*

- o Ensure quotes are complete and accurate, including mark-ups, labor, materials, and credits.
- o Avoid submitting change orders with errors to maintain trust and credibility.
- o Be prepared to answer questions and justify costs with backup documentation.

d) *Managing Turnover and Handover*

i) *Start early and plan thoroughly*

- o Begin turnover planning early—ideally from project kickoff.
- o Develop a detailed closeout strategy and turnover process that aligns with owner expectations.
- o Use tools like "Right Finish" meetings and turnover workshops to align all stakeholders.

ii) *Set clear expectations*

- o Define deliverables, timelines, and responsibilities well in advance.
- o Ensure owners, vendors, trade partners, and facilities teams understand what to expect.
- o Clarify move-in readiness, including furniture, tech setup, and cleaning.

iii) *Communicate continuously*
   - Maintain open, proactive communication with the owner throughout the project.
   - Use frequent check-ins to manage expectations and avoid surprises.
   - Overcommunicate when issues arise—"swarm the problem" to resolve it quickly.

iv) *Be flexible and responsive*
   - Acknowledge that issues will occur—respond quickly and collaboratively.
   - Be ready to adjust plans based on real-time conditions and the owner's needs.
   - Keep focusing on solutions, not blame.

v) *Align with end users*
   - Understand the needs of end users and facilities managers—not just the owner's administrative and project management team.
   - Use commissioning, QA/QC, and prepunch processes to ensure readiness.
   - Help guide owners through the turnover process, especially if they're unfamiliar with construction.

e) *Managing Acceleration and Unrealistic Expectations*

i) *Set expectations early*
   - Establish a realistic, contingency-inclusive schedule at project start.
   - Use tools like RightStart meetings to align on goals and constraints.
   - Clearly define the difference between "wants" and "bought scope."

ii) *Communicate openly and often*
   - Have honest, upfront conversations with the owner about what's achievable.
   - Use weekly OAC and schedule review meetings to track progress and reset expectations.
   - Share schedule updates and material logs regularly to avoid surprises.

iii) *Plan for acceleration scenarios*
   - Develop a contingency plan for potential acceleration needs.
   - If acceleration is requested, analyze feasibility and cost impacts before committing.
   - Present options with associated premiums so the owner can make informed decisions.

iv) *Collaborating with subcontractors*
   - Don't commit to acceleration without buy-in from trades.
   - Walk the site with subs and owners to review logistics and constraints.
   - Ensure field-verified input before making promises.

v) *Avoid overpromising*
   - Never overpromise and underdeliver—it damages trust and project outcomes.
   - If unsure, pause and consult the team before responding to owner requests.
   - Be transparent about external risks (e.g., weather, supply chain) and how they're being managed.

f) *Managing Owner Payment Delays*

i) *Establish clear payment processes early*
   - Define payment terms, submission requirements, and approval workflows at project start.
   - Hold a kickoff meeting to align expectations and responsibilities.
   - Clarify who the gatekeepers are (e.g., owner reps, architects, finance teams).

ii) *Communicate and follow up consistently*
   - Discuss payment status in every OAC or billing meeting.
   - Provide accurate, complete backup documentation to avoid delays.
   - Follow up politely but persistently when payments are late—"poke" when needed.

iii) *Understand the root cause*
   - Investigate why the delay occurred—is it procedural, personnel-related, or due to missing info?
   - Use that insight to prevent future delays or adjust expectations with subcontractors.

iv) *Be transparent with subcontractors*
   - Keep subs informed about payment timelines and any known delays.
   - Align expectations to avoid cash flow issues or work stoppages.
   - Emphasize the importance of accurate billing to maintain trust with the owner.

v) *Co-create and enforce accountability*
   - Develop a co-owned payment process with all stakeholders.
   - Use group pay app reviews to streamline approvals and reduce turnaround time.
   - Hold all parties accountable to the agreed-upon process—no pay app should sit idle.

2) *Best Practices for Contractor-Related Dispute Flashpoints*

a) *Managing Liquidated Damages*

i) *Prevention starts at contracting*
   - Discuss LDs early during bidding and contract negotiation.
   - Ensure LD clauses are clearly defined, realistic, and balanced—consider pairing with incentive clauses.

- o Avoid contracts with LDs when possible, or negotiate their removal via change order.

ii) *Set and manage the schedule proactively*

- o Establish a baseline schedule with float and contingencies.
- o Use realistic planning, preconstruction coordination, and continuous schedule tracking.
- o Drive the schedule from day one and adjust early when delays are identified.

iii) *Communicate and document delays*

- o Notify contractors immediately when delays occur and document thoroughly.
- o Use delay letters, recovery plans, and owner meeting updates to keep all parties informed.
- o Track delays and their causes to defend against unjust LD assessments.

iv) *Hold contractors accountable*

- o Make sure subs are aware of LD risks and notify them formally if they're causing delays.
- o Provide opportunities for correction and supplement labor if necessary to stay on track.
- o Keep LDs visible throughout the project to maintain urgency and focus.

v) *Collaborate and course correct*

- o Use weekly schedule reviews and coordination meetings to stay aligned.
- o Be transparent with owners and subs about schedule risks and mitigation strategies.
- o If LDs become a threat, act quickly to "douse the fire" with acceleration, shift work, or resource reallocation.

b) *Managing Progress Payments and Preventing Misuse*

i) *Establish clear payment processes*

- o Define payment terms, documentation requirements, and billing cycles at project start.
- o Use a detailed schedule of values to track billing against actual progress.
- o Communicate expectations clearly with subcontractors and owners.

ii) *Align billing with actual progress*

- o Ensure pay apps reflect real, verified work in place—not projections or padded values.
- o Involve superintendents or field staff in reviewing progress and validating pay apps.

- o Require proof for stored materials and ensure proper documentation is submitted.

iii) *Implement financial safeguards*

- o Use lien waivers, sworn statements, and certified payroll to protect against misuse.
- o Consider joint checks or retention strategies for financially risky subcontractors.
- o Prequalify subs to ensure financial stability and billing integrity.

iv) *Monitor and review regularly*

- o Conduct routine reviews of pay applications with all stakeholders.
- o Address discrepancies or overbilling immediately and directly with the subcontractor.
- o Document all conversations and corrective actions to maintain accountability.

v) *Communicate and educate*

- o Educate all parties on billing expectations and approval procedures.
- o Keep the owner informed about how billing is managed and verified.
- o Foster a culture of transparency and accountability from the start.

c) *Addressing Contractor Character and Reputation Concerns*

i) *Prevent through prequalification*

- o Conduct thorough prequalification before bidding, including reference checks with past clients and lower-tier subcontractors.
- o Require key personnel (e.g., foremen) to attend prebid and postbid meetings to ensure alignment from the start.
- o Be selective, especially in sensitive markets like healthcare or government work.

ii) *Address concerns professionally*

- o If issues arise, have direct, candid conversations with the contractor to clarify expectations and outline corrective actions.
- o Avoid airing internal issues in front of the owner—present a united front to maintain team credibility.
- o Focus on actions and performance, not hearsay or personal judgments.

iii) *Monitor and document performance*

- o Keep a regular feedback loop on contractor performance throughout the project.
- o Document all interactions and issues to support future decisions or disputes.
- o Use trust-but-verify strategies to ensure accountability without micromanaging.

iv) *Communicate and collaborate*
   - If character concerns surface mid-project, restructure communication paths to reduce friction and maintain team cohesion.
   - Overcommunicate expectations and progress to avoid misunderstandings.
   - Work through detailed execution plans and maintain regular check-ins to guide performance.

v) *Protect your reputation*
   - Maintain professionalism and integrity—do what you say you will do.
   - Avoid personal or subjective commentary; focus on facts and outcomes.
   - Remember that reputation is earned over time—both yours and theirs.

d) *Managing Technical Inexperience in Contractors*

i) *Prevent through rigorous prequalification*
   - Vet contractors thoroughly during bidding and postbid interviews.
   - Speak with past clients and lower-tier subs to assess real-world performance.
   - Avoid hiring contractors who lack the specific experience for the project scope.

ii) *Set clear expectations early*
   - Clearly define roles, responsibilities, and performance standards at project start.
   - Ensure the right team members are assigned to the right tasks.
   - Require the actual field leadership (e.g., foreman) to be present during scope reviews.

iii) *Support through mentorship and training*
   - If already under contract, offer coaching, training, or mentorship where appropriate.
   - Assign internal resources or subject matter experts to assist with technical gaps.
   - Encourage a willingness to learn and improve, especially if the contractor is receptive.

iv) *Act quickly when issues arise*
   - Address performance concerns immediately and directly.
   - Develop a remediation plan with clear corrective actions and timelines.
   - If necessary, replace or supplement the contractor to protect project outcomes.

v) *Document and communicate*
   - Keep a record of issues, conversations, and corrective steps.

- Maintain open communication with the contractor and internal team to avoid escalation.
- Focus on solutions, not blame, and protect the integrity of the project team.

e) *Managing Underbidding and Estimate Accuracy Issues*

i) *Prevent through thorough pre-bid and post-bid processes*
- Conduct detailed prebid and postbid interviews to vet the contractor's understanding and scope coverage.
- Compare bids against industry benchmarks and competitor pricing to identify red flags.
- Require detailed, broken-out scopes of work (SOWs) and schedules of values (SOVs).

ii) *Clarify scope and expectations early*
- Ensure clear, complete, and consistent bid documents to reduce ambiguity.
- Engage field teams and installers during estimating to capture real-world insights.
- Discuss owner expectations and local labor conditions to align pricing with reality.

iii) *Analyze and address discrepancies immediately*
- If underbidding is discovered, review the estimate in detail and identify gaps.
- Have honest conversations with the contractor about their options and responsibilities.
- Use factual data and documentation to support discussions and avoid disputes.

iv) *Support with oversight and QA/QC*
- Implement internal QA/QC reviews of estimates before submission.
- Learn from past mistakes by analyzing previous underbids and adjusting future practices.
- Encourage value engineering (VE) discussions to manage overages constructively.

v) *Communicate transparently with stakeholders*
- Be upfront with owners and design teams if issues arise—transparency builds trust.
- Offer solutions, not just problems, to maintain project momentum.
- Document all communications and decisions to protect against future claims.

f) *Managing Contractor Schedule and Productivity Issues*

i) *Plan early and collaboratively*

- Involve contractors in schedule development to ensure buy-in and accuracy.
- Use tools like pull planning, LEAN scheduling, and manpower-loaded schedules.
- Include baseline schedules in bid documents to set clear expectations from the start.

ii) *Monitor progress continuously*

- Conduct daily huddles, weekly progress meetings, and 3-week lookaheads.
- Track actual production against the schedule and identify slippage early.
- Use software tools and visual workflows to stay organized and transparent.

iii) *Communicate and coordinate*

- Make the schedule a standing topic in all subcontractor meetings.
- Foster open collaboration to identify roadblocks and resolve them quickly.
- Ensure field teams and PMs are aligned on priorities and constraints.

iv) *Take swift corrective action*

- When delays occur, implement a recovery plan immediately.
- Use tools like 72-hour notices or corrective action plans to prompt accountability.
- Be willing to adjust sequencing, add manpower, or reallocate resources to stay on track.

v) *Design for productivity*

- Create a site layout that supports efficient workflows.
- Remove barriers to productivity by addressing material, manpower, or access issues.
- Build schedule contingencies and workarounds to absorb minor disruptions.

g) *Managing Resources and Balancing Issues*

i) *Plan early and set clear expectations*

- Define resource needs (labor, materials, equipment) during bidding and postbid interviews.
- Communicate schedule expectations and delivery timelines upfront.
- Include resource planning and procurement strategies in preconstruction planning.

ii) *Monitor and track continuously*

- Use procurement logs, manpower tracking, and equipment schedules to monitor availability and performance.

- Compare actual progress versus planned durations to identify shortfalls early.
- Track and forecast delays in labor or material delivery to stay ahead of issues.

iii) *Respond quickly and collaboratively*

- When issues arise, implement corrective actions immediately, such as 72-hour notices or recovery plans.
- Be prepared to supplement labor, adjust sequencing, or source alternate materials.
- Work with subcontractors to co-create solutions and maintain momentum.

iv) *Build flexibility into the plan*

- Establish contingency plans for labor shortages, material delays, or equipment failures.
- Design workflows with batch sizes and flow efficiency to maintain productivity even with limited resources.
- Stay nimble and adaptable—especially in today's volatile construction environment.

v) *Communicate impacts and solutions*

- Keep owners and stakeholders informed of resource-related risks and mitigation strategies.
- Share financial and schedule impacts transparently to build trust and avoid disputes.
- Use data-driven insights to support decisions and maintain accountability.

h) *Managing Union or Nonunion Labor Disputes*

i) *Plan ahead and understand the market*

- Know the local labor environment and project-specific requirements (e.g., prevailing wage, project labor agreements [PLA], national maintenance agreements [NMA]).
- Discuss labor expectations early with the owner and bidders, especially during prebid and postbid phases.
- Be prepared for dual gate systems or picketing on projects with mixed labor forces.

ii) *Maintain a neutral, performance-focused approach*

- Emphasize that the goal is quality, value, and productivity, regardless of labor affiliation.
- Avoid drawing attention to union status unless it becomes a project issue.
- Focus on contractual obligations and performance, not affiliations.

iii) *Communicate proactively and respectfully*
  - Hold prejob meetings with organized labor to set expectations and avoid misunderstandings.
  - Keep lines of communication open with all labor groups to address concerns before they escalate.
  - Avoid inflammatory language or actions that could trigger labor disputes.

iv) *Respond appropriately if issues arise*
  - If disputes occur, implement dual gate systems or other site logistics to maintain operations.
  - Address disruptions quickly and work with labor representatives to resolve tensions.
  - Keep the focus on project continuity and team cohesion.

v) *Align with project type and delivery method*
  - Recognize that public and private projects may have different labor requirements.
  - Ensure all parties understand what is legally required and contractually allowed.

i) *Managing Work Quality Disputes*

i) *Set clear quality expectations early*
  - Use preinstallation meetings, mockups, and first-work-in-place reviews to establish quality benchmarks.
  - Review specifications and standards with all stakeholders, including special inspectors.
  - Ensure all trades understand the expected level of finish and performance before work begins.

ii) *Implement a strong QA/QC program*
  - Develop and follow a formal QC plan throughout the project.
  - Use tools like Procore Observations or similar platforms to track and manage deficiencies.
  - Assign a dedicated quality manager or designate team members to oversee QA/QC efforts.

iii) *Monitor and correct continuously*
  - Conduct real-time quality checks during installation—not just at the punch list.
  - Address issues immediately upon discovery to avoid rework and delays.
  - Differentiate between completion lists and punch lists to manage expectations and timing.

iv) *Communicate and document*
   - Keep subcontractors informed of deficiencies and provide clear corrective action plans.
   - Use daily huddles, coordination meetings, and system reviews to stay aligned.
   - Document all quality-related discussions and resolutions to support accountability.

v) *Prevent rework through planning*
   - Avoid rework by getting it right the first time—through training, supervision, and clear communication.
   - Reference mockups or early inspections to align all parties on what "done right" looks like.
   - Reinforce a culture of pride in workmanship and ownership of quality.

j) *Managing Commissioning-Related Disputes*

i) *Start early and plan thoroughly*
   - Begin the commissioning (Cx) process during the design and submittal review phases.
   - Develop a detailed commissioning plan that outlines responsibilities, timelines, and expectations.
   - Ensure the CxA, contractors, and equipment vendors are aligned from the start.

ii) *Define roles and expectations clearly*
   - Clarify who is responsible for each commissioning task and ensure all parties understand the process.
   - Use kickoff meetings, regular Cx meetings, and checklists to maintain alignment.
   - Establish how deficiencies will be tracked and resolved before testing begins.

iii) *Maintain open communication*
   - Hold regular meetings with the CxA and subcontractors to review progress and resolve issues.
   - Invite the CxA for job walks to identify potential issues before formal testing.
   - Communicate ball-in-court responsibilities clearly to avoid delays and finger-pointing.

iv) *Be proactive and responsive*
   - Address issues in real time with supporting documentation and collaborative problem-solving.
   - Assign a team member to manage precommissioning checklists and expedite issue resolution.
   - Ensure resources are available during testing to fix issues on the spot.

v) *Align with quality and turnover goals*
   - Integrate commissioning with the QA/QC process to catch issues early.
   - Treat commissioning as a final validation of quality, not just a checklist.
   - Understand that bugs and issues are part of the process—what matters is how quickly and effectively they're resolved.

k) *Managing Safety-Related Disputes*

i) *Prioritize prevention through planning*
   - Develop a site-specific safety plan early and communicate it clearly to all contractors.
   - Include safety expectations, EMR/DART thresholds, and compliance requirements in prequalification and bid documents.
   - Conduct pretask planning, toolbox talks, and safety audits regularly.

ii) *Foster a culture of safety ownership*
   - Emphasize that safety is everyone's responsibility, from management to tradespeople.
   - Encourage open reporting of hazards and near-misses without fear of retaliation.
   - Gain buy-in from individual workers to ensure adherence to safety protocols.

iii) *Monitor and enforce consistently*
   - Use daily check-ins, safety standdowns, and PSAs to reinforce expectations.
   - Assign dedicated safety personnel to oversee compliance and correct unsafe behavior.
   - Avoid overreliance on removal—focus on education and accountability.

iv) *Respond swiftly and transparently to incidents*
   - When accidents occur, stop work immediately, investigate thoroughly, and communicate findings.
   - Use tools like 5 Whys or root cause analysis to identify and correct systemic issues.
   - Share lessons learned and update safety plans to prevent recurrence.

v) *Document and communicate*
   - Maintain accurate safety records, including premobilization paperwork and daily reports.
   - Ensure all parties are aware of incident response protocols and reporting structures.
   - Communicate remedial actions and follow-ups clearly to all stakeholders.

l) *Managing and Preventing Mechanics Lien Disputes*
   i) *Prioritize prevention through financial controls*
      - Ensure executed contracts and lien-related clauses are in place before work begins.
      - Require and track sworn statements, lien waivers, and notices of commencement (NOC).
      - Avoid hiring contractors with a history of lien filings by vetting financial reliability during prequalification.
   ii) *Maintain timely and transparent payment practices*
      - Address payment issues promptly to avoid escalation.
      - Keep change orders and billing cycles current to prevent misunderstandings.
      - Communicate with owners and subs about payment status and delays to maintain trust.
   iii) *Monitor and document continuously*
      - Use change logs and payment tracking tools to ensure all work and costs are accounted for.
      - Make direct calls to key suppliers if needed to verify payment status and avoid surprises.
      - Document all communications and agreements to support dispute resolution if needed.
   iv) *Act quickly if a lien threatens*
      - If a lien is threatened, convene an "all hands" meeting to resolve the issue collaboratively.
      - Consider paying the disputed amount to release the lien while holding back funds to recover costs later.
      - Treat lien threats as a sign that something in the process is broken and fix it immediately.
   v) *Foster a culture of proactive dispute resolution*
      - Encourage field teams to resolve payment concerns early, before legal action is taken.
      - Recognize that liens are often a last resort—most can be avoided with early communication and fair treatment.
      - Use lien threats as an opportunity to review and improve internal processes.

3) *Best Practices for Miscellaneous-Related Dispute Flashpoints*
   a) *Managing Schedule and Coordination Disputes*
      i) *Plan early and set clear expectations*
         - Communicate schedule expectations during bidding and subcontracting.

- o Understand subcontractor availability and commitments before awarding contracts.
- o Include specialty vendors and tiered subs in early planning and coordination efforts.

ii) *Build collaborative and flexible schedules*

- o Use pull planning, lookahead schedules, and visual work area maps to align all trades.
- o Involve subcontractors in schedule development to ensure buy-in and feasibility.
- o Be ready to resequence work or shift trades to noncritical areas when conflicts arise.

iii) *Maintain constant communication*

- o Hold daily huddles, weekly progress meetings, and monthly coordination sessions.
- o Use meetings to review upcoming work, identify conflicts, and adjust plans.
- o Encourage open dialogue between trades to resolve issues before they escalate.

iv) *Track and adjust in real time*

- o Monitor material deliveries, manpower levels, and equipment availability.
- o Quickly implement temporary solutions when delays or conflicts occur.
- o Use field-level coordination between superintendents and foremen to stay agile.

v) *Reinforce accountability*

- o Set the tone that contractors are obligated to meet the project's schedule, not their own.
- o Document expectations and hold trades accountable for delays or missed commitments.
- o Emphasize that coordination is a shared responsibility for project success.

b) *Managing Document Discrepancies*

i) *Prioritize early and thorough review*

- o Conduct constructability reviews during preconstruction to catch inconsistencies.
- o Use page-turn sessions with the owner, design team, and field experts to align on scope and intent.
- o Review every note, detail, and specification to ensure clarity before bidding.

ii) *Develop clear and detailed scopes*
   - Create comprehensive scopes of work that bridge gaps in the documents.
   - Use overlays and comparison tools to track changes between drawing sets.
   - Establish a contract review checklist to ensure consistency and completeness.

iii) *Communicate and collaborate proactively*
   - Engage with the owner and A/E early when discrepancies are identified.
   - Use RFIs, submittals, and coordination meetings to clarify and resolve issues before work begins.
   - Involve trade contractors and field teams in identifying and addressing unclear or conflicting details.

iv) *Act quickly and document everything*
   - When discrepancies arise, respond promptly with written direction to avoid rework or delays.
   - Document all findings, decisions, and communications to support future claims or negotiations.
   - Establish a formal process for identifying, reviewing, and resolving discrepancies.

v) *Accept That Not All Issues Can Be Prevented*
   - Recognize that some discrepancies will surface during construction.
   - Focus on early detection and collaborative resolution to minimize impact.
   - Use the experience to improve future document review and coordination processes.

c) *Managing Risk and Reward Sharing Disputes*

i) *Establish Terms Early and Clearly*
   - Define risk and reward structures during contract negotiation, before bidding or GMP is established.
   - Clarify how contingency, shared savings, and performance incentives will be handled.
   - Use contract language to document expectations, responsibilities, and dispute resolution processes.

ii) *Align incentives with accountability*
   - Ensure that those who assume risk are eligible for reward, and vice versa.
   - Avoid rewarding parties who offload risk or underperform, while recognizing those who go above and beyond.

- o Consider performance-based incentives tied to schedule, cost savings, or quality outcomes.

iii) *Promote transparency and communication*
- o Maintain ongoing conversations about risk and reward throughout the project.
- o Notify all parties promptly when risk-sharing conditions are triggered.
- o Use change logs and tracking tools to monitor shared risks and benefits in real time.

iv) *Be fair and flexible*
- o Recognize that shared risk/reward models can foster teamwork when structured fairly.
- o Be open to adjusting terms if unforeseen issues arise that impact the original balance.
- o Avoid creating adversarial conditions—collaboration should be the goal.

v) *Avoid ambiguity*
- o Prevent disputes by clearly defining what happens in scenarios like delays, scope changes, or cost overruns.
- o Ensure all parties understand who owns what risk and how rewards will be distributed.

d) *Managing Disputes with Litigious Participants*

i) *Identify and understand early*
- o Use prequalification and background research to identify participants with a history of litigious behavior.
- o Understand their motivations and triggers to better anticipate potential disputes.

ii) *Set the tone with clear communication*
- o Establish a collaborative tone early in the project to reduce adversarial tendencies.
- o Address concerns directly and calmly—don't mirror aggressive behavior.
- o Focus on mutual goals and partnership benefits to shift the mindset from conflict to cooperation.

iii) *Stay proactive and document everything*
- o Maintain detailed daily reports, RFIs, and correspondence to protect your position.
- o Be prepared in the background but avoid escalating unless necessary.
- o Use contract language and processes to guide responses and decisions.

iv) *De-escalate and redirect*
- o If litigious behavior surfaces, engage leadership or legal teams early to manage risk.

- o Redirect conversations to specific issues and solutions, not threats or posturing.
- o Consider mediation or structured resolution processes if disputes persist.

v) *Avoid fueling the fire*

- o Don't overreact to threats of litigation—many are bluffs.
- o Avoid public or emotional responses that could undermine team morale or trust.
- o Focus on resolving the root issue, not rhetoric.

e) *Managing Communication-Related Disputes*

i) *Communicate clearly and proactively*

- o Use simple, direct language and avoid ambiguity.
- o Be upfront and transparent about intentions, expectations, and decisions.
- o Ensure everyone who needs the information is included in the conversation.

ii) *Document everything*

- o Follow up verbal conversations with written confirmation (e.g., email or RFI).
- o Use documentation to clarify agreements, decisions, and responsibilities.
- o Remember: "If it's not in writing, it didn't happen."

iii) *Establish communication protocols early*

- o Set expectations in kickoff meetings and project startup sessions.
- o Define who communicates what, when, and how (e.g., phone, email, meetings).
- o Create a communication plan with the owner, architect, and subcontractors.

iv) *Address miscommunication promptly*

- o Don't let misunderstandings linger—reset and clarify immediately.
- o Use in-person meetings when possible to resolve complex or sensitive issues.
- o Be calm and constructive—avoid blame and focus on resolution.

v) *Build a culture of openness and respect*

- o Make team members feel heard and valued—don't dismiss concerns.
- o Encourage frequent, direct conversations and regular updates.
- o Promote trust and collaboration through consistent engagement.

f) Managing Disputes Related to Unique Project Conditions

i) *Identify and plan for unique conditions early*

- o Conduct site reconnaissance, soil studies, and logistical assessments during preconstruction.

- o Use prebid walkthroughs and RightStart meetings to highlight site-specific challenges.
- o Develop a project execution plan that addresses known risks and constraints.

ii) *Communicate expectations upfront*

- o Be transparent with all stakeholders—don't downplay difficult conditions.
- o Clearly describe site limitations, access issues, and environmental factors in the scopes of work and bid documents.
- o Align expectations with owners, designers, and subcontractors before work begins.

iii) *Customize scope and logistics*

- o Tailor logistics plans, temporary conditions, and resource allocations to the site's unique needs.
- o Include allowances or contingencies for difficult conditions in bids and contracts.
- o Quantify and communicate potential impacts to cost and schedule early.

iv) *Choose the right partners*

- o Select subcontractors and vendors who are experienced with similar project types or conditions.
- o Discuss challenges during prebid and postbid interviews to ensure alignment and capability.

v) *Stay flexible and adaptive*

- o Be ready to adjust plans and strategies as conditions evolve.
- o Maintain a problem-solving mindset and collaborate with the team to overcome obstacles.
- o When conditions are less than ideal, offer creative solutions or trade-offs to maintain momentum.

g) *Managing RFI-Related Disputes*

i) *Prevent through document quality and preconstruction reviews*

- o Conduct thorough constructability reviews during schematic and design development phases.
- o Involve trade partners and field experts early to identify gaps before documents are finalized.
- o Ensure clear, complete, and coordinated documents to reduce the need for RFIs.

ii) *Set expectations and processes early*

- o Establish a clear RFI protocol during preconstruction, including format, content, and response timelines.

- o Discuss RFI expectations with subcontractors and the design team before construction begins.
- o Use a contractual or agreed-upon RFI workflow to manage volume and prioritization.

iii) *Filter and manage RFI submissions*
- o Have the CM/GC team review RFIs before submission to ensure they are necessary and well-formed.
- o Encourage subs to combine related questions and avoid submitting RFIs for information already available.
- o Use emails or field clarifications for minor issues that don't impact cost or schedule.

iv) *Collaborate with the design team*
- o Maintain open communication with the A/E team to expedite responses and reduce bottlenecks.
- o Hold regular meetings to review outstanding RFIs, prioritize critical items, and track impacts.
- o Use the owner's influence when necessary to ensure timely responses from the design team.

v) *Monitor and adjust*
- o If the RFI rate becomes excessive, analyze the root cause—document quality, unclear scope, or subcontractor behavior.
- o Adjust strategies accordingly, whether through additional training, clarification, or escalation.

h) Managing Disputes from Poorly Detailed Novel Structures

i) *Prioritize early detection*
- o Conduct thorough constructability reviews during preconstruction to catch unclear or missing details.
- o Use postbid interviews and shop drawing reviews to surface and clarify ambiguous design elements.
- o Encourage early engagement with subs and designers to flag potential issues before they impact the schedule.

ii) *Clarify design intent collaboratively*
- o When details are unclear, infer intent based on experience and propose practical solutions.
- o Submit RFIs promptly to document and resolve discrepancies.
- o Schedule coordination meetings with A/E and subs to align on expectations and finalize details.

iii) *Communicate expectations clearly*
- o Set clear expectations for quality and final product appearance based on contract documents.

- Ensure that design revisions and updates are issued well in advance of field execution.
- Use mockups or visual references when necessary to align on outcomes.

iv) *Maintain flexibility and team harmony*
- Focus on team-based problem solving rather than blame.
- Be proactive in resolving issues as they arise, minimizing delays and rework.
- Establish a repeatable protocol for addressing detailing issues to streamline future responses.

v) *Document and track resolutions*
- Keep a record of design clarifications, RFI responses, and field adjustments.
- Use these records to inform future projects and improve detailingy standards.

i) Managing Disputes from Unforeseen Circumstances

i) *Plan for the unplannable*
- While unforeseen conditions can't be predicted, they can be anticipated with contingency planning.
- Include allowances, schedule buffers, and investigative tools (e.g., GPR, soil borings) in preconstruction.
- Build flexibility into project execution plans, especially for renovations or complex sites.

ii) *Respond quickly and decisively*
- When an unforeseen issue arises, act immediately to assess the situation and prevent further impact.
- Gather data, define the problem, and develop a resolution plan with input from field teams and experts.
- Use a "swarm the issue" approach—mobilize resources and decision-makers quickly.

iii) *Communicate transparently*
- Notify the owner, architect, and key stakeholders as soon as the issue is identified.
- Be clear about the "what," "why," and "how"—what happened, why it matters, and how it will be resolved.
- Maintain open, honest communication to build trust and avoid escalation.

iv) *Collaborate on solutions*
- Work as a team to identify the best path forward, balancing cost, schedule, and quality.

- o Use rough order of magnitude (ROM) estimates and scenario planning to help owners make informed decisions.
- o Document the issue and resolution process for future reference and learning.

v) *Learn and adapt*
- o After resolution, analyze the root cause and determine if similar issues can be prevented in the future.
- o Update project protocols and contingency strategies based on lessons learned.

j) *Managing Weather-Related Disputes*

i) *Proactive weather planning*
- o Analyze historical and seasonal weather trends to anticipate delays.
- o Integrate weather considerations into early project planning stages.
- o Treat weather as a manageable risk, not an unforeseen event.

ii) *Realistic and flexible scheduling*
- o Avoid assuming 100% productivity every day.
- o Build float and contingency time into the project schedule.
- o Adjust timelines based on seasonal expectations and regional climate.

iii) *Budgeting and resource allocation*
- o Include slush funds for winter/wet conditions.
- o Allocate for temporary protection (e.g., enclosures, drainage systems).
- o Ensure bids reflect realistic weather-related costs and allowances.

iv) *Mitigation and recovery strategies*
- o Develop workarounds and backup plans for weather-impacted trades.
- o Use overtime or premium time to recover critical path delays.
- o Act quickly to implement mitigation (e.g., dousing leaks, resequencing tasks).

v) *Accountability and communication*
- o Document weather delays responsibly and consistently.
- o Reserve claims for extreme or prolonged weather events only.
- o Communicate early with stakeholders to negotiate time or cost adjustments.

k) *Managing Events*

i) *Establish clear contractual language*
- o Include an events clause in all contracts during negotiations.
- o Clearly define what qualifies (e.g., natural disasters, pandemics, government shutdowns).
- o Understand the limitations of insurance (e.g., builder's risk) and cost recovery options.

ii) *Communicate early and transparently*
- o Notify the full project team immediately when an event occurs.

- o Share updates regularly to maintain alignment and reduce confusion.
- o Use formal documentation (e.g., letters) to outline impacts and constraints.

iii) *Be flexible and adaptive*

- o Recognize that no one can control events—adaptability is key.
- o Be willing to adjust schedules, scopes, and expectations as needed.
- o Use past experiences (e.g., COVID-19) to inform response strategies.

iv) *Assess and mitigate quickly*

- o Evaluate the impact of the event as soon as possible.
- o Identify temporary alternatives or workarounds to keep progress moving.
- o Implement mitigation steps early to prevent further disruption.

v) *Foster a collaborative mindset*

- o Approach the situation with a "no one wins" mentality—focus on fairness.
- o Educate owners and stakeholders on the nature and implications of having collaborative teams.
- o Work together to find equitable solutions and maintain project momentum.

l) *Managing Team Dynamics and Project Culture*

i) *Build teams intentionally*

- o Understand team members' strengths and weaknesses when forming teams.
- o Be deliberate about team composition to ensure compatibility and balance.
- o Use tools like a responsibility matrix to clarify roles and reduce friction.

ii) *Set expectations early*

- o Establish clear roles, responsibilities, and communication norms from day one.
- o Conduct partnering meetings and kickoff events to align on values and goals.
- o Define a check-in cadence to monitor team health and address concerns early.

iii) *Foster open communication*

- o Promote a culture of transparency and psychological safety.
- o Ensure open-door policies are in place for all team members and stakeholders.
- o Encourage regular feedback loops to surface and resolve issues quickly.

iv) *Address issues immediately*
   - Don't let conflict fester—intervene early to prevent escalation.
   - Be willing to replace team members who consistently disrupt team cohesion.
   - Use a "douse early" approach to resolve problems before they spread.

v) *Invest in culture and relationships*
   - Organize teambuilding activities and offsite engagements to strengthen bonds.
   - Recognize and reward collaboration and shared success.
   - Unite the team around a common purpose and reinforce it throughout the project.

10.6 Limitations and Future Research

While this chapter offers valuable insights into PDM through the D-PAD framework, several limitations should be acknowledged. The Delphi study was conducted with a relatively small and regionally concentrated sample of professionals from a single CM firm in the Midwest United States, which may limit the generalizability of the findings. Additionally, the study focused primarily on leadership-level participants, potentially overlooking perspectives from field staff, subcontractors, and other stakeholders who often experience disputes more directly. The reliance on expert opinion, while insightful, introduces subjectivity, and the absence of fully integrated case studies limits the practical illustration of the D-PAD framework. Furthermore, the framework itself, while structured, may not fully capture the dynamic and overlapping nature of real-world disputes.

Future research should aim to broaden the participant base to include more diverse roles, regions, and project types, enhancing the applicability of findings across the industry. Longitudinal studies could provide empirical evidence on the long-term effectiveness of PDM strategies. There is also a need to explore how digital tools such as BIM, AI-driven analytics, and project management platforms can support or enhance proactive dispute practices. Comparative studies across industries may yield transferable strategies, while deeper exploration into behavioral and cultural factors could enrich the human-centered dimensions of PDM. Finally, developing quantitative metrics to evaluate the success of prevention, avoidance, and dousing strategies would strengthen the framework's practical utility and support its integration into standard project delivery processes.

10.7 Conclusion

This chapter reinforces the critical role of PDM in addressing the multifaceted nature of conflicts in construction and related industries. By applying the D-PAD framework, industry professionals can anticipate and mitigate disputes before they escalate. The findings highlight that many disputes stem from predictable flashpoints, such as unclear documentation, unrealistic expectations, and coordination failures. Through early planning, transparent communication, and collaborative problem-solving, these issues can often be prevented or avoided entirely. Even when disputes arise unexpectedly, a proactive mindset enables teams to respond swiftly and constructively. Ultimately, PDM empowers stakeholders to foster a culture of foresight, accountability, and continuous improvement—leading to more resilient project outcomes and stronger industry relationships.

This chapter has also demonstrated the value of the Delphi method in advancing PDM within the construction industry. By engaging experienced professionals across multiple roles, the study identified critical dispute flashpoints and aligned them with the D-PAD framework—prevention, avoidance, and dousing. The findings underscore that while many disputes are preventable through early planning, communication, and QC, others require adaptive responses and collaborative resolution. The expert consensus offers actionable insights and best practices that can be integrated into project delivery strategies to reduce conflict, enhance team dynamics, and improve project outcomes. As the industry continues to evolve, embracing proactive, data-informed approaches like Delphi will be essential for fostering a culture of foresight and resilience in dispute management.

References and Suggested Further Reading

Ameyaw, E.E., Hu, Y., Shan, M. et al. (2016). Application of Delphi method in construction engineering and management research: a quantitative perspective. *Journal of Civil Engineering and Management* 22 (8): 991–1000. https://doi.org/10.3846/13923730.2014.945953.

Beiderbeck, D., Frevel, N., von der Gracht, H.A. et al. (2021). Preparing, conducting, and analyzing Delphi surveys: cross-disciplinary practices, new directions, and advancements. *MethodsX* 8: 101401. https://doi.org/10.1016/j.mex.2021.101401.

Gutierrez-Bucheli, L., Goh, J.T., Rashidi, A. et al. (2024). Adopting immersive technologies in construction training: determining educational decision-making criteria through a Delphi technique. *Smart and Sustainable Built Environment*. https://doi.org/10.1108/SASBE-08-2023-0202.

Hallowell, M.R. and Gambatese, J.A. (2010). Qualitative research: application of the Delphi method to CEM research. *Journal of Construction Engineering and Management* 136 (1): 99–107. https://doi.org/10.1061/(ASCE)CO.1943-7862.0000137.

Kermanshachi, S., Rouhanizadeh, B., and Dao, B. (2020). Application of Delphi method in identifying, ranking, and weighting project complexity indicators for construction projects. *Journal of Legal Affairs and Dispute Resolution in Engineering and Construction* 12 (1): https://doi.org/10.1061/(ASCE)LA.1943-4170.0000338, https://doi.org/10.1061/(ASCE)LA.1943-4170.0000338.

Khodyakov, D., Grant, S., Kroger, J., and Bauman, M. (2023). *RAND Methodological Guidance for Conducting and Critically Appraising Delphi Panels*. RAND.

Linstone, H.A. and Turoff, M. (2002). *The Delphi Method - Techniques and Applications*. Addison-Wesley Publishing Company. https://doi.org/10.2307/1268751.

Mollaoglu, S., Sparkling, A., and Thomas, S. (2015). An inquiry to move an underutilized best practice forward: barriers to partnering in the architecture, engineering, and construction industry. *Project Management Journal* 46 (1): 69–83. https://doi.org/10.1002/pmj.21469.

Naji, K.K., Gunduz, M., and Mansour, M.M. (2024). Development of an integrated hybrid risk assessment system for construction disputes during the preconstruction phase using the Delphi method. *Journal of Construction Engineering and Management* 150 (7): https://doi.org/10.1061/JCEMD4.COENG-14492.

Discussion

1) The Delphi panel agrees that PDM requires dedicated resources, but smaller firms express concerns about cost and feasibility.
   Prompt: How can PDM principles be scaled for smaller organizations without compromising effectiveness? What practical steps can be taken to balance cost and proactive dispute prevention?
2) Consensus from the Delphi study suggests that PDM should be embedded in organizational policy, but some participants argue that project-level implementation is more practical.
   Prompt: Debate the merits of institutionalizing PDM at the organizational level versus applying it on a project-by-project basis. Which approach offers greater long-term benefits?
3) Experts in the Delphi study identify cultural differences among project teams as a major dispute trigger.
   Prompt: How can PDM strategies be adapted to address cultural and organizational diversity? What role does early stakeholder engagement play in mitigating these risks?

11

Dispute Risk Analysis (DRA)

Readers' Learning Objectives and Outcomes

1) *Identify various actors embodying flashpoints and triggers of conflicts and disputes in the industry.*
2) *Hypothesize scenarios and analytical models of conflicts and disputes in the industry, preparatory to the application of artificial intelligence for predictions.*
3) *Create a primer to match and apply prevention, avoidance, or dousing strategies to each conflict's and dispute's flashpoints and triggers.*

11.1 Introduction

Dispute risk analysis (DRA) is a technique for analyzing, identifying, and assessing factors likely to jeopardize the success of a project or a venture or prevent it from achieving its goals partly or wholly. Hence, DRA in construction has to do with identifying, analyzing, and assessing potential triggers of disputes that are likely to jeopardize the success of a project or the achievement of its goals, and proactively (1) evaluate and deploy applicable internal structures and systems on hand to prevent them; (2) determine how to navigate around and avoid them; and (3) equip the teams with the know-how and approaches to swiftly douse them, before they flare up and escalate to resource-demanding negotiation, mediation, arbitration, or litigation. The "Project Risk Assessment/Management Cycle" below encapsulates and graphically illustrates the process in depth.

DRA can be conducted alone or done alongside value engineering and buildability analysis of a construction project. The DRA searchlight should beam across the project owner and representatives, participating professionals and consultants, contractor and representatives, subcontractors, suppliers or vendors, contract documents, project location, and impacting external factors. A comprehensive DRA precedes attempts at predicting disputes by artificial intelligence tools and other algorithms (Figure 11.1).

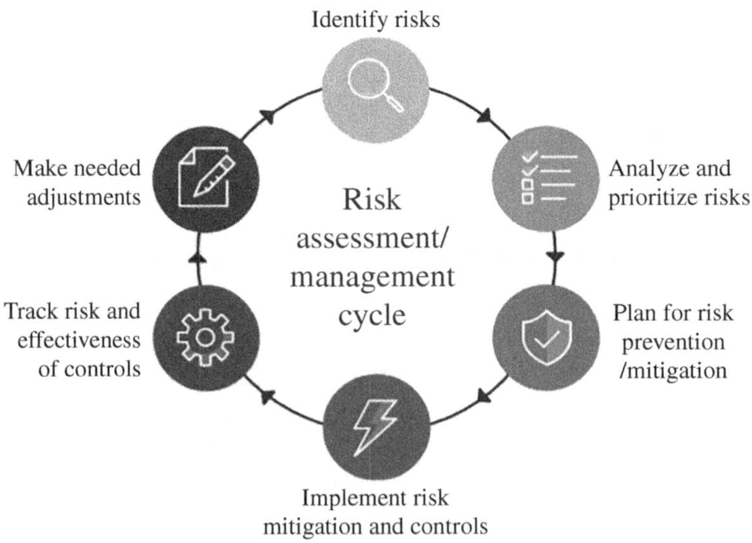

Figure 11.1 Project risk assessment/management cycle. *Source:* Smartsheet Inc/ https://www.smartsheet.com/content/project-risk-assessment?srsltid=AfmBOoqluGC-tu_-G2YnFyzNRTx5S3GNGZ_aR-vYHWKA-WF8ykID_sMC/ last accessed on Jan 21, 2026.

11.1.1 Owner and Representatives

Analyzing project dispute risks from the owners' and representatives' perspectives involves identifying potential issues that could be flashpoints for conflicts and disputes between owners and representatives, and other participants and stakeholders. DRA here needs to be approached in the following sequence:

1) *Examine and understand project requirements and scope*: This involves a comprehensive and thorough examination of the contract documents that include the program, drawings, specifications, addenda, bid instruction, contract, other preconstruction documents, amendments, and instructions. Interrogate and clarify any negligent aspects, such as incomplete drawing details and specifications, discrepancies between or among documents, and any ambiguities in the contract clauses. Ensure that the roles and responsibilities of the parties and participants are clearly defined, delineated, and understood. The parties and participants include owners and representatives, contractors, subcontractors, and suppliers or vendors.
2) *Identify potential flashpoints or triggers of disputes*: The commonest triggers include the following five circumstances:
   a) *Scope changes or creep*: Changes or modifications of the work and scope readily lead to disputes if not appropriately handled or managed. From the onset, there should be a clearly set out process for handling changes (addenda and change orders) with clear protocols for approval and documentation.
   b) *Workmanship and work quality*: The quality of workmanship translates to the overall project or work quality. When the delivery falls below professional and/or reasonable expectations, it can readily lead to disputes. Hence, both should be continually monitored to ensure that they meet standards and contractual requirements.
   c) *Cost overruns*: Projects are often conducted with limited budgets and exceeding them is a ready trigger of disputes. Ensure to work with a realistic budget for the project and establish protocols for handling and managing changes in costs.
   d) *Delays*: Each project has a timeline and exceeding it can be a potential trigger for disputes. Leveraging on forecasting and projections from experience, the project schedule must be established from the onset with full cognizance and consideration of potential bottlenecks that can bring about delays. Such bottlenecks should be proactively evaluated with clarity on how delays from them are to be handled, communicated, and managed.
   e) *Communication breakdown*: Communication is key in delivering a successful project unencumbered with misunderstandings, conflicts, and disputes. Where there's poor communication or a communication breakdown

between the parties, such is a fertile ground for disputes between them. Hence, communication channels must be free, open, and timely between the parties to clear any misunderstandings that can lead to conflicts and disputes.

3) *Identify and assess dispute risks and determine how to manage them:* Proactively and virtually construct and deconstruct the project from start to finish, and identify and assess potential risks associated with financial, technical, and operational aspects. Evaluate risk mitigation mechanisms that can be deployed, such as contingencies, bonds and insurance, as well as specific contractual clauses and provisions.
4) *Tap into experience and evaluate relevant historical data and precedents:* Previous projects, data, and experience accumulated from them are always helpful resources in analyzing and appreciating potential dispute risks of current projects on hand. They provide insights into potential risks to anticipate or predict, and how to better handle and manage them based on how they were navigated around or resolved.
5) *Evaluate stakeholder relationships:* Evaluate actual or possible working relationships with the project owner group (owner and representatives), and between the owner group and the contractor and subcontractors and suppliers. Uneasy or strained relationships, and where there is little collaboration and coordination, are fertile grounds for conflicts and disputes. Hence, this aspect needs to be dissected and shored up if a successful project outcome is to be envisaged.
6) *Examine contractual clauses:* Contractual clauses should be comprehensively examined and combed through with a fine-tooth comb to exterminate ambiguities, and clarify or delineate responsibilities, potential liabilities, and indemnities.
7) *Deploy robust monitoring and reporting system:* Establish a protocol for regular inspections, documentation, and reporting to track project progress, identify, isolate, and handle issues proactively. That is a deployment of the whole gamut of proactive dispute management (PDM), where dispute issues that cannot be prevented are either avoided or doused before they escalate to demand resource-consuming negotiation, mediation, arbitration, or litigation. Track, document, and report any changes of scope, schedule, or budget in order to facilitate the effective application of PDM's dispute prevention, avoidance, and dousing (D-PAD).
8) *Engage PDM's D-PAD:* Deploy PDM's D-PAD as suitable from the onset and through the project delivery process. Some potential dispute issues are better prevented, while others can be better avoided; failing which any uprisings can be doused before they escalate to demand resource-consuming negotiation, mediation, arbitration, or litigation.

9) *Review feedback and continuously improve:* It is possible for the deployment of PDM's D-PAD to not arrest all dispute issues 100% on every project, especially at the first or early start times of their use or application. Hence, to improve their efficacy, postconstruction evaluation, review, and continuous improvement are imperative. One such evaluation, review, and continuous improvement process is the implementation of dispute reputation rating (DRR), which will be discussed in a later chapter.

Following this sequence of approaches enables helpful proactive identification, isolation, and assessment of project dispute risks associated with owners and their representatives, assisting in their better handling and management to ensure successful, mitigated dispute or dispute-free project outcomes.

11.1.2 Participating Professionals and Consultants

In addition to the design professionals and consultants working for project owners, such as architects and civil engineers, others include building services engineers (mechanical, electrical, communication, etc.), structural engineers, surveyors, quantity surveyors and/or cost engineers, program managers, construction or project managers, procurement managers, and estimators. Analyzing project dispute risks from the participating professionals' and consultants' perspectives involves identifying potential issues that could be flashpoints for conflicts and disputes between parties stemming from the contributions of these participants. DRA here needs to be approached similar to that of owners and representatives (examine and understand project requirements and scope; identify potential flashpoints or triggers of disputes; identify and assess dispute risks, and determine how to manage them; tap into experience and evaluate relevant historical data and precedents; evaluate stakeholder relationships; examine contractual clauses; deploy robust monitoring and reporting system; engage PDM's D-PAD; and review feedback and continuously improve), with the following additional considerations:

1) *Understand the roles and responsibilities of participating professionals and consultants*: Understanding participating professionals' and consultants' roles and responsibilities involves a review of their agreements and contracts to ascertain and ensure that their scope of services and expectations are clearly defined and aligned with project needs.
2) *Evaluate communication and coordination protocols*: Evaluate communication protocols to ensure effective interactions among participating professionals, consultants, and other project stakeholders. Poor communication with participating professionals and consultants can lead to misunderstandings that can end in conflicts and disputes. Assess how participating professionals and consultants can effectively and efficiently coordinate with other

project stakeholders to accomplish objectives and successful outcomes. The inefficiencies from poor coordination are potential triggers of conflicts and disputes.

Following this sequence of approaches enables helpful proactive identification, isolation, and assessment of project dispute risks associated with participating professionals and consultants, assisting in their better handling and management to ensure successful, mitigated dispute or dispute-free project outcomes.

11.1.3 Contractors and Representatives

DRA here follows the same sequence as that of owners and representatives; namely, examine and understand project requirements and scope; identify potential flashpoints or triggers of disputes; identify and assess dispute risks, and determine how to manage them; tap into experience and evaluate relevant historical data and precedents; evaluate stakeholder relationships; examine contractual clauses; deploy robust monitoring and reporting system; engage PDM's D-PAD; and review feedback and continuously improve.

11.1.4 Subcontractors

DRA here follows the same sequence as that of owners and representatives; namely, examine and understand project requirements and scope; identify potential flashpoints or triggers of disputes; identify and assess dispute risks, and determine how to manage them; tap into experience and evaluate relevant historical data and precedents; evaluate stakeholder relationships; examine contractual clauses; deploy robust monitoring and reporting system; engage PDM's D-PAD; and review feedback and continuously improve. One of the additional considerations for participating professionals and consultants also applies, namely, evaluate communication and coordination protocols.

11.1.5 Suppliers or Vendors

Just as for subcontractors, DRA here follows the same sequence as that of owners and representatives; namely, examine and understand project requirements and scope; identify potential flashpoints or triggers of disputes; identify and assess dispute risks, and determine how to manage them; tap into experience and evaluate relevant historical data and precedents; evaluate stakeholder relationships; examine contractual clauses; deploy robust monitoring and reporting system; engage PDM's D-PAD; and review feedback and continuously improve. One of the additional considerations for participating professionals and consultants also applies; namely, evaluate communication and coordination protocols.

11.1.6 Contract Documents

DRA in contract documents involves a thorough preconstruction review and quality control of all aspects, from drawings and specifications to addenda and the agreement, to clear all ambiguities and discrepancies. Doing so assists in identifying and clearing any dispute risk issues and PDM. DRA can be conducted on contract documents in the following sequence:

1) *Review work scope and documentation*: Review and ascertain the scope of work to ensure that the contract documents are commensurate with it. In other words, the scope of work must be clearly defined, and its contract documents must be comprehensive and free of negligent omissions, ambiguities, discrepancies, and vagueness that are ready causes of misunderstandings and disputes. The work breakdown must be clearly delineated, and the parties' responsibilities must be clearly defined with overlaps and gaps promptly addressed.
2) *Check for consistency in contract documents and address identified ambiguities and discrepancies*: Where specifications duplicate information that can be extracted from drawings, ensure that there is consistency and no discrepancies. Contract clauses related to change orders and scope modification must have clear procedures, including how they are to be documented, priced, and approved. Payment terms and procedures must be clearly spelled out to be strictly followed. Clear procedures and provisions must be made for handling delays and constructive acceleration, time extensions, and early delivery, as well as any applicable penalties and bonuses. Similar considerations apply to guarantees and warranties.
3) *Assess allocation of responsibilities and risks*: Assess how risks and responsibilities are allocated to the parties to ensure equity and reasonable distribution. Procedures and provisions for handling and managing unforeseen circumstances and challenges must be clearly spelled out, understood, and accepted by the parties. Requirements for bonding and insurance must be reviewed to ascertain adequate coverage of potential risks and liabilities.
4) *Examine key contract clauses*: Key contract clauses to examine and be very clear on include those to do with bonds and insurance, change orders and scope modifications, schedule of rates if applicable, payment terms, completion dates, and deadlines, as well as guarantees and warranties. It must be ensured that all the necessary bonds and insurance are taken out by the parties, with clarity on how they would be applied whenever necessary. Clarity must bear on the process of handling scope modifications and associated change orders, including how they are documented, priced, and approved. Where the work is known, but its extent or content is indeterminable, an

agreed schedule of rates of values must be used. The procedures for handling payments must be unambiguous, including how to invoice and request them, approve, and pay them. Review the project's milestones and deadlines relative to the final completion timeline, and be clear on the provisions for handling delays, constructive acceleration, time extension, bonuses, and penalties. Review guarantee and warranty clauses to be clear on the parties' obligations in relation to performance and quality, and how issues are to be handled when they arise.

5) *Review compliance and performance clauses in the contract*: Compliance and performance clauses have to do with quality and standards, as well as inspections and testing. Review the clauses to ensure that quality and standards are unambiguously specified, especially where alternates are allowed with "equality" qualifications, i.e., an alternative product or installation claimed to be equal to the one specified must be ascertained to be exactly or acceptably so. Ensure clarity on the procedures for inspections and testing, including which party is to conduct them and how issues are to be addressed, if any.
6) *Analyze contractual obligations, bonuses, and penalties*: Clarify all contractual obligations of the parties and associated deliverables and performance requirements, as well as any bonuses or penalties pertaining to subpar performance. The obligations must be determinably realistic or achievable. Ascertain the bonuses or penalties and damages to follow subpar or nonperformance or delays, and ensure that they are fair and equitable.
7) *Review historical and dispute precedents*: Review past similar contracts and project types and the disputes witnessed with them to identify potential common issues and risks to anticipate. Critically analyze the legal precedents to appreciate how similar issues and risks were handled and resolved, and project how to feed that into the PDM of the current project and its contract.
8) *Evaluate communication and documentation procedures*: Assess and understand the procedures of communication between the parties and among project participants and stakeholders, including how notices, changes, and approvals are communicated and documented. Ensure the clarity of the requirements for documenting all aspects of the project process, including changes, approvals, performance deliverables, and issues.
9) *Engage PDM's D-PAD*: Configure how to deploy and integrate PDM's D-PAD into the administration of the current project with its contract and participants. If strategies are nonexistent to prevent identified dispute risks, conceive new and proven strategies to navigate around and avoid latent or potential project disputes, and measures to swiftly douse them if they show signs of flaring up.
10) *Monitor and review continuously*: As the project delivery advances, continually monitor and review latent or potential disputes to timely identify and

proactively address emerging risks and issues either by avoidance or dousing, whichever is more appropriate to apply. Gather feedback to serve as a bank of historical resources to refer to for PDM of future projects and their contracts, together with their participants and stakeholders.

By critically and systematically analyzing construction contract documents with these outline steps, you can identify and proactively manage dispute risks and their impacts on project delivery, outcomes, and relationships.

11.1.7 Location

Analyzing dispute risks from a construction project's location involves understanding how the physical site and its attributes can trigger potential conflicts and disputes. Factors such as site conditions, environmental and regulatory requirements, differences in people and cultures, and accessibility may all pose potential risks that can lead to disputes. The following are ways to appropriately analyze, consider, and accommodate or allow for these risks to ensure a smooth or smoother project execution, outcomes, and relationships:

1) *Analyze site conditions*: Site conditions must be assessed, which can portend dispute risks, including geotechnical conditions of the site, the topography, and any existing structures, infrastructures, or services. Geotechnical issues that pose dispute risks can stem from soil conditions, underlying rock structures, and groundwater levels. Poor soil conditions and unanticipated subterranean issues can lead to contested, unbudgeted costs that can trigger disputes between the parties. Some topographical conditions, such as the nature of slopes, elevations, and surface water runoff, may call for modifications and unbudgeted additional work and costs that can trigger disputes between the parties. Issues associated with demolition, relocation, and reconstruction of existing structures and services can be dispute triggers between the parties due to the extra costs.
2) *Assess prevalent environmental and regulatory factors*: Prevalent environmental and regulatory factors that must be considered that can portend dispute risks include environmental impact assessments or reports related to protected habitat, flooding, siltation, and pollution control; local, state, and federal regulatory compliance requirements, such as zoning ordinances, building codes, health and safety regulations; applicable permits, and approvals. If all these are not comprehensively proactively considered, managed, and allowed for, they can lead to issues capable of triggering disputes between the parties.
3) *Evaluate site accessibility and logistics*: As poor access to the site can lead to material and equipment delays that can trigger disputes, it is crucial to assess the site's accessibility and logistics. Evaluate the site layout to determine staging areas for construction activities. Difficult and tight site layout can make

staging more problematic, leading to subsequent operational inefficiencies that can lead to tensions, conflicts, and disputes. Site safety and security measures should also be evaluated, as unsafe and insecure conditions can lead to accidents, theft, and other vices that can result in losses and disputes.

4) *Examine other site-specific risks*: Examine and make provisions or accommodations for other site-specific risks, such as the prevailing local weather conditions that can affect construction activities. Seasonal changes and extreme weather conditions can disrupt construction and lead to delays and extra costs that are certain triggers of disputes between the parties. Proactively identifying and devising measures to mitigate the impacts of the site's potential natural hazards, such as flooding, earthquakes, and heavy storms, can help to prevent, avoid, or douse possible disputes.
5) *Review historical and community factors*: Research the site for any past encumbrances and issues related to previous land uses, including construction. Previous historical site issues and potential community opposition, if existent, pose concerns that can lead to new construction disputes. Identifying and devising measures to deal with or overcome the site's historical issues and community factors can help to prevent, avoid, or douse possible new construction disputes.
6) *Analyze impact on project cost and schedule*: Evaluate how site conditions and requirements may affect project cost and schedule. Unexpected site conditions, requirements, and associated delays can lead to cost and time extensions and overruns that can trigger conflicts and disputes.
7) *Assess contractual provisions related to site conditions*: Thoroughly review contractual clauses on site conditions, such as provisions for unforeseen site conditions and adjustments for site-related issues. Ensure that these clauses are clear, unambiguous, and adequately address associated potential risks. Such clauses dictate how dispute risks from unforeseen site conditions and requirements will be handled and proactively managed.
8) *Engage in open communication and documentation*: Timely document detailed records of site conditions, issues, and requirements, and promptly communicate with all parties and stakeholders. Such documentation and communication will help in managing expectations and preventing or avoiding conflicts and disputes.
9) *Engage PDM's D-PAD*: Deploy and engage the three tripods of PDM: D-PAD in relation to site conditions, requirements, and potential issues. This can be accomplished through the conduct of thorough site surveys and assessments to identify potential site issues and device mitigation strategies before project construction commencement. To further proactively manage potential dispute risks from unforeseen or unexpected site conditions, requirements,

and issues, as well as regulatory changes, contingency planning for them is key. Continuous monitoring of site conditions as the project is ongoing is also helpful in identifying and addressing site-specific issues and timely deescalating and dousing them before they trigger conflicts and disputes.

By critically and systematically analyzing project location and site-specific conditions, requirements and issues following these outline steps, you can identify and proactively manage dispute risks stemming from them and their impacts on project delivery, outcomes, and relationships.

11.1.8 External Factors

Analyzing dispute risks from external factors of a project involves understanding and managing realities outside and beyond the control of the project team that can still significantly impact the outcome. Factors such as legal, economic, environmental, and social influences can pose potential risks that can lead to disputes. The following are ways to appropriately analyze, consider, and proactively accommodate or allow for these risks to ensure a smoother project execution, outcomes, and relationships:

1) *Identity and understand legal and regulatory compliance risks*: Assess the project's compliance with applicable municipal, state, and federal regulations. Changes in laws and regulations can impact the project and approval process, leading to delays that are triggers of disputes. Delays and issues with permits can lead to project delays and disputes. Ensure that contract terms allow for and accommodate these uncertainties.
2) *Analyze economic and financial risks*: Analyze current market conditions and trends, as fluctuations in prices and availability can lead to disputes over changes and cost overruns. Ascertaining the economic stability of the region or country of project location is important in analyzing and proactively managing potential dispute risks from this factor. Economic downturn can affect project financing and the solvency of all project stakeholders, including material suppliers and vendors. Hence, it is vital to understand the financial health of project stakeholders, including owners and investors. Financial difficulties can lead to disputes over timely periodic or interim payments and timely project completion.
3) *Assess environmental and climate risks*: Noncompliance with environmental regulations and requirements can attract fines and project stoppage; for instance, waste collection and disposal, emissions, and conservation measures. Hence, environmental regulations and requirements applicable to the project must be timely ascertained and strictly complied with. Review and conduct

environmental impact assessments on the project, and note any potential risks posed. Unanticipated environmental issues can lead to additional costs and delays that are both potential triggers of disputes. Local climate and prevalent weather impact construction activities. Extreme weather events, where they occur, can affect the project timeline from delays and damages, for instance, heavy rainfall, hurricanes, and severe storms. Associated delays and damages are ready triggers of disputes.

4) *Analyze social and community factors*: Ascertain any potential community concerns and opposition to the project. Protests from opposing groups and negative community sentiments and actions can lead to project delays and disputes. Hence, assessing and understanding the local social impacts of the project is crucial: for instance, the project's impact on residents and local businesses, increased traffic, noise, and disruptions, all of which can lead to complaints, protests, and disputes.
5) *Review political and legal environments*: Just as legal environments, political instability and changes of government and policies can affect project execution and lead to disputes. For instance, a project might have started under the Biden administration without tariffs on important components but is proceeding under the Trump administration with significant tariffs that are affecting both costs and supply chains, and subsequently, the project's costs and timelines. Ensure that contract terms allow for and accommodate these uncertainties, but the importance of forecasting cannot be overstated in this scenario.
6) *Consider supply chain and logistical risks*: Timely identify and accommodate or contain any risks in the project's supply chain, as well as any delays in material delivery or issues with suppliers. Disruptions in the supply chain, such as from sudden and shifting tariffs, can lead to project delays, higher costs, and disputes. Assess and navigate around any logistical challenges, such as transportation infrastructure and site access, as these can affect project timelines and costs that trigger disputes.
7) *Evaluate safety and health risks*: Noncompliance with safety and health regulations can lead to unmitigated incidents and accidents, legal claims, and work stoppages. Hence, timely ascertaining and complying with all applicable project's safety and health regulations are crucial. Safety- and health-related crises and issues can dictate workforce availability and project schedules. A major example of a health crises that affect all projects and work sectors was the COVID-19 pandemic. Those that failed to manage this crisis properly with renegotiation of project terms ended in fiercely contested disputes.

Critically and systematically analyzing these external factors by following these outline steps, devising risk mitigation strategies, continuously reviewing and monitoring them, you can identify and proactively manage dispute risks stemming from them, as well as their impacts on project delivery, outcomes, and relationships.

11.2 Conclusion

A cornerstone of PDM in construction and related industries is DRA. This chapter emphasizes that disputes often stem from predictable flashpoints such as unclear documentation, unrealistic expectations, and coordination failures. By conducting thorough risk assessments before project commencement, stakeholders can identify these triggers early and implement preventive measures. DRA supports the integration of PDM principles—D-PAD into project planning and execution, enabling teams to anticipate risks and mitigate them before they escalate. Ultimately, applying DRA fosters a culture of foresight, accountability, and collaboration, reducing costly disruptions and strengthening relationships among project participants. It transforms dispute handling from a reactive process into a proactive strategy, ensuring more resilient and successful project outcomes.

Chapter Review Questions

1) Dispute risk analysis (DRA) is only applicable after a project begins. True/False?
2) DRA helps match potential disputes to appropriate PDM strategies like D-PAD. True/False?
3) Evaluating subcontractors and suppliers is unnecessary in DRA if they've worked with the contractor before. True/False?
4) External factors such as political and environmental influences are considered in DRA. True/False?
5) DRA should be conducted only by external consultants. True/False?
6) Communication and coordination protocols among professionals are part of DRA evaluation. True/False?
7) DRA is a reactive tool used to resolve disputes after they occur. True/False?
8) Which of the following is a key goal of dispute risk analysis (DRA)?
   a) To delay the project start
   b) To identify and proactively manage potential dispute triggers
   c) To eliminate the need for contracts
   d) To increase litigation opportunities

9) What does the D-PAD acronym stand for in PDM?
   a) Design, planning, analysis, delivery
   b) Dispute prevention, avoidance, dousing
   c) Documentation, planning, arbitration, delivery
   d) Delay, payment, agreement, dispute
10) Which of the following is NOT typically analyzed during DRA?
   a) Site conditions
   b) Project team roles
   c) Weather forecasts for the next decade
   d) Communication protocols
11) Which group is responsible for collecting data during DRA?
   a) External auditors
   b) Lead PDM and internal team members
   c) Legal counsel only
   d) Subcontractors
12) Which of the following is considered an external factor in DRA?
   a) Subcontractor experience
   b) Project budget
   c) Regulatory and legal influences
   d) Site layout
13) What is the benefit of having long-term relationships with suppliers in DRA?
   a) Reduced need for contracts
   b) Increased dispute risk
   c) Better predictability of dispute behavior
   d) Elimination of quality checks
14) Which of the following should be reviewed to understand professional roles in DRA?
   a) Weather reports
   b) Social media profiles
   c) Contracts and agreements
   d) Marketing materials
15) Dispute Risk Analysis (DRA) is used to identify and manage ______ dispute triggers.
16) The DRA process involves matching risks to the PDM framework known as ______.
17) Evaluating the ______ and responsibilities of professionals helps prevent role ambiguity.
18) Site conditions, logistics, and accessibility are part of the ______ analysis in DRA.

19) External factors such as climate and politics are outside the control of the ______ team.
20) DRA should be conducted during the ______ stage to proactively address risks.

Exercises

Exercise 11.1 DRA Matrix Development for a New Project

Objective:
Create a customized DRA matrix for a new construction project.
Instructions:
- Choose a hypothetical or real project (e.g., a university science building).
- Identify at least five categories of potential dispute triggers (e.g., owner expectations, subcontractor coordination, site conditions).
- For each category, list:
  - Specific risk factors
  - Likelihood of occurrence (low/medium/high)
  - Impact severity (low/medium/high)
  - Recommended D-PAD strategy (prevention, avoidance, dousing)
- Compile the information into a DRA matrix.

Outcome:
Learn how to systematically assess and prioritize dispute risks before project execution.

Exercise 11.2 DRA Audit of a Completed Project

Objective:
Retrospectively apply DRA to a completed project to identify missed risks and lessons learned.
Instructions:
- Select a completed project (real or fictional) that experienced disputes (e.g., delays due to site conditions or subcontractor issues).
- Conduct a mock DRA audit by:
  - Identifying dispute flashpoints that were not anticipated
  - Mapping those flashpoints to D-PAD strategies that could have been used
  - Recommending improvements for future DRA implementation
- Present your audit findings in a short report or team discussion.

Outcome:
Understand how DRA could have proactively mitigated disputes and how to improve future risk analysis.

Discussion

1) How can early DRA shape the culture of a construction project team? Discuss how identifying risks before they occur can influence communication, accountability, and collaboration throughout the project lifecycle.
2) DRA focuses on conflict-related risks. How can DRA be integrated with broader project risk management tools (e.g., safety, financial, environmental)? Should DRA be a standalone process or part of a unified risk strategy?

12

AI-Driven Framework for Proactive Dispute Management in Construction and Related Industries

Readers' Learning Objectives and Outcomes

1) *Understand the nature and causes of disputes in the construction industry, including their classifications.*
2) *Identify dispute management and resolution methods from traditional approaches to digitalized solutions.*
3) *Develop an understanding of artificial intelligence (AI), including its core technologies, concepts, models, and methodology.*
4) *Explore the role of AI in dispute management, including the development of predictive models and an AI-driven framework to proactively manage disputes.*

12.1 Introduction

In the architecture, engineering, and construction (AEC) industry, claims arise frequently, often resulting in disputes. A claim can be defined as "a request for compensation for damages incurred by any party to the contract and can refer to either a time extension or money reimbursement." Then, a dispute arises when one party makes a claim that is subsequently rejected by the other (Kalogeraki and Antoniou 2024). The complex and competitive nature of construction projects has led to frequent conflicts and disputes (El-Sayegh et al. 2020). In addition, the involvement of a high monetary value of construction projects is often a reason for disputes. Construction contracts are prone to disputes because they govern complex, multiparty relationships. Moreover, the inherent uncertainty and complexity of the industry often hinder effective communication and information exchange among project teams. With numerous stakeholders involved and multiple project stages from inception to completion, disputes become difficult to resolve and are therefore common.

Frequent disputes have led to substantial cost overruns and delays while also damaging working relationships among project participants, ultimately hindering the success of construction projects (Wang J et al. 2023). Similarly, Francis et al. (2022) indicated that disputes adversely affect the project budget, the schedule, and the quality, thereby affecting the project performance at large. In the worst-case scenario, disputes lead to bankruptcy of the contractor and termination of construction projects. Therefore, disputes need to be well managed to minimize their effects (Saseendran et al. 2020). Litigation and alternative dispute resolution (ADR) methods are commonly used to resolve disputes after they materialize. However, these methods are often costly in terms of time and money and require extensive documentation, which demands high levels of skill and effort. Disputes need to be resolved proactively and settled early, as their inextricability will aggregate negative impacts on performance outcomes. Thus, the construction industry, along with academic researchers, has been exploring various strategies and methodologies to prevent disputes before they arise. A proactive approach is considered more effective than a reactive one.

In recent years, several innovative technologies have been coupled with existing dispute management practices to achieve proactive dispute management, which predominantly include blockchain, smart contracts, building information modeling (BIM), and artificial intelligence (AI). For example, a blockchain-based claim management using Hyperledger Fabric is proposed to improve the quality and reliability of records and enhance the availability of relevant information for effective claim and dispute management (Torkanfar et al. 2023). "A new construction-specific framework, Decentralized Construction Enabling Transparent Resolution (DCENTR), has been developed using blockchain technology to address construction disputes with greater transparency and significant reductions in effort, time, and cost" (Saygili et al. 2022). In addition, BIM can be utilized to manage construction disputes more effectively, highlighting the correlations between the primary causes of disputes and the benefits of BIM adoption (Wang, J et al. 2023). The BIM-dispute resolution (BIM-Dres) framework aligns legal aspects and finalized contractual provisions with the phases of a construction project while also identifying the key stakeholders involved or impacted by these elements (Olatunji 2016). In recent years, several studies have been initiated to utilize various AI technologies like machine learning (ML), deep learning (DL), NLP, and large language models (LLMs) to predict and manage disputes. Specifically, AI technologies are being developed to predict the probability of disputes, dispute resolution methods, causes of disputes, and litigation outcomes by training AI algorithms using digital construction data. Therefore, this chapter aims to develop an AI-driven framework for proactive dispute management in the construction industry.

12.2 Nature of Disputes in the Construction Industry

12.2.1 Classifications of Disputes and Their Causes

A comprehensive literature review was carried out among 43 selected journal articles that primarily investigate the causes of disputes to identify the major types of disputes. Francis et al. (2025) developed a typology of causes of dispute that includes 12 major types of disputes. The causes were further refined, leaving a total of 87 causes for classification. The causes were carefully examined, and similar causes were grouped together. Eight main dispute categories were thereby identified, which include payment and finance-related, delay- and time-related, design- and scope-related, contract-related, behavioral- and people-related, site-related, quality-related, and external factors-related disputes. Table 12.1 depicts the main dispute types and the related causes of disputes.

As depicted in Table 12.1, the causes are grouped and formed into eight major dispute classifications. Among these groups, contract-related disputes account for the majority of causes (17). It seems that there are many problems that occur and turn into several disputes due to issues in contract documents. A poorly written contract causes disputes between the parties, as the project is based on a weak agreement, leading to miscommunication between the parties. Especially in lump-sum contracts, since the owner and the contractor agree on a fixed price, it is very important to have complete drawings and specifications before the contractor is selected. However, clients mostly do not have skilled employees who are qualified to prepare these documents. Contractors, on the other hand, stated

Table 12.1 Major types of disputes and number of related causes.

Main dispute category	Causes of disputes	Number of causes
Payment and finance related	Payment issues	12
	Project exceeds budget	
	Unrealistic pricing in bids	
	Contractor's cash flow difficulties	
	Client's cash flow difficulties and project funding issues	
	Variation in material prices	
	Inaccurate project cost estimation	
	Delay in evaluation of completed works and issuing payment certificate	
	Inappropriateness and disagreement in payment terms and modalities	
	Bankruptcy of contractor	
	Prolonging the financial cost	
	Back-to-back payment mechanism	
Delay and time related	Progress delay by contractor/subcontractor/nominated subcontractor/supplier	14
	Delay in completion of project	
	Late decision-making/response/instruction/approval by the client	
	Late instruction/decision/information/approval/inspection by the consultant	
	Delay in handing over the site to the main contractor	
	Delay in supplying material and equipment	
	Extension of time (EOT) claim issues	
	Governmental and third-party approval delays	
	Acceleration of work	
	Delay in mobilization by contractor	
	Underestimation of project duration by the contractor	
	Weak schedule practices by the contractor	
	Lack of progress reporting	
	Poor time estimation of the project by the engineer	

Table 12.1 (Continued)

Main dispute category	Causes of disputes	Number of causes
Design and scope related	Scope of work changes	08
	Design errors and poor-quality design	
	Frequent design revisions	
	Complexity of the construction projects	
	Variation in quantities	
	Incomplete employer requirements	
	Inadequate site investigation	
	Uncertain or continually changing performance requirements	
Contract related	Ambiguities and contradictions in contractual documents	17
	Misinterpretation of contractual terms and conditions	
	Incomplete contract document	
	Unfair risk allocation	
	Errors in contract documents	
	Breach of contract	
	Compliance issues	
	Lack of dispute resolution methods and strategies	
	Suspension of work	
	Termination of contract	
	Poor contract management	
	Inadequate risk identification/allocation	
	Selection of contractors by low bid	
	Inefficient bidding, inadequate investigation, and preparation	
	Unjust and untimely presentation of claims	
	Lack of construction legislation	
	Unattended/unsettled claims	
Site related	Differing site conditions	05
	Poor supervision and site management	
	Lack of collaboration with various project trades	
	Contractors' lack of project staff and workers	
	Poor management of the construction process and resources, and shortage of resources	

(Continued)

Table 12.1 (Continued)

Main dispute category	Causes of disputes	Number of causes
Quality related	Poor quality of work	04
	Defects in completed work	
	Poor-quality material	
	Rework	
People and behavior related	Lack of communication	14
	Client's attitude	
	Administration issues	
	Opportunistic and adversarial behavior of the contractor	
	Lack of experience/knowledge/skills	
	Incompetent contractor	
	Technical inadequacy of contractor	
	Lack of trust and misunderstanding among parties	
	Client as a layman/inexperienced	
	Lack of teamwork	
	Conflict of interest in both parties	
	Reluctance of project participants to deal with changes	
	Unawareness of the governing rules/local laws of the related governmental agencies	
	Negligence/lack of professionalism	
External factors	Labor and workforce issues	13
	Uncontrolled factors (weather, disasters)	
	Material shortage	
	Price escalation	
	Changes in government regulations	
	Fluctuation of foreign exchange rates	
	Changes in market conditions and environmental regulations	
	Political and bureaucratic challenges	
	Uncertainty of the construction projects	
	Environmental issues/pollution	
	Force majeure events	
	Legal and economic factors	
	Fluctuation of tax rate	

that the errors in contract documents cause serious problems and unresolved claims among parties that end up in complex disputes.

A total of 14 factors have been classified under delay and time-related disputes. Project delays occur in several ways, such as designer time delay, time delay from the main contractor, time delay from the owner, work stoppage, and time delay from the subcontractor. Throughout the project, there were numerous changes to the scope of work and challenges in adhering to the specified design requirements. These factors led to significant modifications in the design, resulting in delays in finalizing the drawings and plans. A study conducted by Alrasheed et al. (2023) in Kuwait revealed that the total value of claims across all disputes amounted to KD 39,661,503 (approximately USD $132 million). Of this total, 51.09% was attributed to disputes arising from delays caused by the main contractor. Additionally, claims related to project delays accounted for 66.90% of the overall claim value. These findings highlight project delays as the most critical factor contributing to disputes in both public and private sector projects. The high proportion of delay-related claims highlights broader issues with the reliability of Kuwait's project delivery system.

A total of 13 factors have been identified as external factors that contribute to disputes in construction projects. Extreme weather conditions and natural disasters, like floods or earthquakes, can disrupt construction schedules, damage completed works, and lead to delays and disputes. Price escalation can significantly impact project budgets and profitability, potentially leading to disputes between project stakeholders. Contractors may seek compensation for increased costs through contract provisions, such as price escalation clauses, while clients may dispute the extent of the cost increases or the applicability of such clauses. The government regulations for obtaining approval vary in complexity and time requirements, depending on jurisdiction. Often, government regulations follow bureaucratic procedures for authorization of the approval required to proceed, thus causing unpredictable delays (El-Sayegh et al. 2020). Further, the authors stated that modifying legislation and regulations can also lead to disputes. Each country has its own laws and rules that are subject to change. These changes might cause disputes during the construction process. Currency fluctuations can affect project costs, especially for projects involving international transactions, leading to disputes over pricing adjustments. Another 14 causes have been identified under people- and behavior-related disputes, which stem from interpersonal dynamics and individual behavior.

Then, payment-related disputes occur due to 12 significant causes. The most common issue that had arisen in the majority of disputes was a cost-related payment claim. Parties disagree either on the payment amount or the payment schedule (Senaratne and Farhan 2023). Delays in payments can significantly hinder construction progress, often leading to slow progress toward project completion. Consequently, delayed payments were identified as the primary source of disputes in both government and private sector projects that led to litigation. These payment delays are particularly prevalent in progress payments. Such payment delays primarily occur due to clients' attitude, incomplete source of funding, and the consultant delaying the evaluation of completed works.

Variations refer to changes in the scope of work made by a party to the contract that consequently affect the project cost. Such variations are common in construction projects. Frequent changes to the scope of work often result in excessive design modifications, leading to delays in completing drawings and plans. Scope changes often arise at the request of the client, especially when the client lacks familiarity with building construction. In such cases, project requirements may not be clearly reflected in the design, leading to frequent disputes related to scope and design changes. Change orders commonly result from an inaccurate bill of quantities (BOQ), poor coordination among project team members, and design errors. These changes can significantly affect the overall cost and timely completion of the project. Five causes were identified under site-related disputes, which mainly arise from factors directly associated with site operations. Quality-related disputes occur due to factors linked to the quality of work, such as poor workmanship, defects in completed work, the use of substandard materials, and rework. The next section describes dispute management practices, ranging from traditional approaches to those used in the digital construction era.

12.2.2 Dispute Resolution

A dispute is a successive phase of conflict that necessitates the resolution of legitimate issues. When a dispute remains unresolved, it proceeds successively from its initial resolution stage: negotiation, mediation, dispute review board (DRB), adjudication, and arbitration, collectively referred to as ADR. Negotiation is an ADR method where parties discuss their grievances and seek early-stage resolution without external intervention (Khan et al. 2025). Negotiations often occur in advance of any potential court order between disputants, as they are the lowest-cost way of reconciling aspects of a disagreement in terms of time, money, and emotional energy involved. The goal of negotiation is to reach an agreement through debate and compromise; it is the primary strategy to resolve disputes among many ADR practices that may avoid court, saving costs, time, and improving outcomes in a less hostile environment for parties. Negotiation is a nonbinding method, whereas arbitration and litigation are binding and formal methods; there, decision-makers are more influential than parties, and decisions are binding on the parties under dispute.

Mediation is recognized as an important alternative form of dispute resolution for all types of disputes (Gregory-Stevens et al. 2016). Mediation is a collaborative and voluntary method where a mediator is chosen to aid disputing parties in reaching a mutual agreement. In mediation, the mediator chosen is neutral and facilitates the interactions among the parties (Kukah et al. 2025). Mediation has been championed as an alternative, supplement, or precedent to adjudication (Gregory-Stevens et al. 2016).

The DRB involves a panel of experts who are independent. These experts are set at the beginning of the project to facilitate the management of disputes that will arise and to provide solutions and decisions (Kukah et al. 2025). DRB and the dispute adjudication board (DAB) are the two variants of dispute boards, and they share similar characteristics and procedures. That is, when a dispute is referred to the dispute board, the board arranges for hearings and the submission of statements by the parties. It issues a fair conclusion, which can be in the form of a nonbinding recommendation in the case of a DRB or a binding decision in the case of a DAB (Abdul-Malak and Senan 2020).

Adjudication is a quick, temporary binding resolution often used in the construction industry. It is an alternative to arbitration for its cost-effectiveness and speed (Khan et al. 2025). The adjudicator or adjudication board can be named in the contract at the time of contract signature, nominated by the parties after the dispute event, or nominated by an adjudication nominating body. A standard construction contract typically requires that adjudication be used to resolve disputes, and if unresolved, the matter may then be escalated to arbitration.

Arbitration in construction disputes is one of the most common ADR techniques worldwide because experts typically issue awards, and these awards are final and binding to the parties (Zand Pazandi et al. 2024). It is considered to be a speedier and more cost-effective way of dispute resolution than litigation (Moza and Paul 2017). In practice, there will either be a sole arbitrator or a three-arbitrator tribunal. In the case of a sole arbitrator, both parties must mutually agree on the arbitrator, or an arbitration institution will appoint one. For a three-arbitrator tribunal, each party typically selects one arbitrator, and the third, who serves as the presiding arbitrator, is usually appointed by a neutral appointing authority.

In the context of construction projects, disputes generally progress through three main phases before potentially resulting in litigation, namely: claim formation, claim rejection, and failure of amicable resolution (Jagannathan and Delhi 2020). Litigation is often chosen when parties require a legally enforceable decision. However, it is generally the least-favored option due to its shortcomings in maintaining confidentiality, preserving business relationships, and delivering timely outcomes. In such cases, arbitration best serves the needs of the parties, offering a more private, flexible, and relationship-friendly method of dispute resolution (Zand Pazandi et al. 2024).

12.2.3 Dispute Management: From the Traditional Approaches to Digitalized Construction

Traditionally, construction disputes were resolved through litigation in the courts. Modern approaches to resolving commercial conflicts have increasingly shifted toward ADR methods. This shift is largely due to the fact that court proceedings are often time-consuming, costly, and can damage relationships between the involved parties. As a result, litigation is now generally considered a last resort for resolving disputes (Faraji et al. 2021). However, any resolution method requires highly specialized skills and proof documents, and disputes are highly technical, involving legal aspects (Cappiello and Carullo 2021). These situations encourage the construction industry to step toward dispute avoidance as per the proverb "prevention is better than cure."

The avoidance of disputes has been addressed over the years at both the industry-wide and project-specific levels. At the industry level, significant efforts have been made to promote a cultural shift from an adversarial approach to a more dispute-averse mindset. The construction industry has been consistently urged to adopt modern management practices, such as partnering and alliancing, with a strong emphasis on involving key stakeholders like clients, contractors, and building users early in the decision-making process. One of the most effective strategies for dispute avoidance is the development and implementation of a communication system that brings all parties onto a single platform, allowing for easy access to project information and regular updates. Additionally, strategies such as fair risk allocation, early nonbinding neutral evaluation, and collaborative partnering play a critical role in effectively managing and minimizing disputes.

Several causal models have been developed in recent years to manage disputes in the construction industry. For example, Ahmed and El-adaway (2023) determined the key associations among the bidding stage–related causes that can lead to potential disputes in order to avoid disputes in bidding stages using network analysis (NA), spectral clustering, and association rule analysis (ARA). Naji et al. (2020) highlight the application of fuzzy logic combined with structural equation modeling (SEM) as a recognized and valid tool for modeling in construction projects. This approach provides a suitable framework for dispute modeling and evaluation. The authors concluded by presenting a hybrid fuzzy-SEM model designed to quantify the probability of dispute occurrence in construction projects. This model enables project stakeholders to predict, identify, and effectively manage potential disputes during the pre-construction phase.

In recent years, the fast digital transformation in the construction industry has stimulated broad research on technology applications in construction management, including dispute management. Technologies such as BIM, blockchain, smart contracts, and AI have begun to be applied in dispute management.

For example, Wang, J et al. (2023) developed a conceptual framework to illustrate how BIM can be adopted to facilitate dispute management throughout the entire lifecycle of construction projects. The framework suggests that issues such as design errors, delays, and change orders can be significantly reduced through the effective use of BIM. Among the various benefits of BIM, improved visual management, enhanced information management, and strengthened collaboration are identified as the most frequently adopted features that contribute to resolving the majority of dispute causes. Tao et al. (2021) mentioned that blockchain technology ensures the authenticity and traceability of project data without the need for intermediaries. This capability helps reduce the time and costs associated with providing data evidence during disputes. On a different note, Torkanfar et al. (2023) proposed a blockchain-based claim management system using Hyperledger Fabric to enhance the quality and reliability of project records. Their system aims to improve the availability of relevant information for effective claim management and to address current challenges in the construction industry, such as the lack of trust, poor collaboration, and limited access to reliable information for dispute resolution. Moreover, the implementation of smart contracts in construction can effectively address issues of nonpayment and delayed payments. By automating and enforcing payment terms, smart contracts help reduce payment-related disputes between clients and contractors, as well as between contractors and subcontractors.

Electronic document management systems (EDMS) play a vital role in streamlining document distribution, accelerating communication, enhancing overall productivity, and improving the quality of documentation—all of which contribute to more efficient dispute management in construction projects. Commonly used data management platforms in the construction industry include Aconex, ProjectWise 365, Autodesk BIM 360, Mi-Doc, Jira, Viewpoint, Speedbrick, Microsoft SharePoint, Proficient, Bluebeam, BuildSpace, and Trello.

Barnett and Treleaven (2018) proposed that integrating blockchain technology with AI can significantly accelerate the resolution of disputes, reduce litigation costs, and provide greater certainty for parties involved in complex and challenging disputes worldwide. In recent years, several AI technologies, including ML, DL, and NLP, have been involved in predicting disputes. For example, Un et al. (2024) developed an innovative model that integrates project characteristics, root causes, and underlying factors to forecast dispute outcomes. The model was tested using a dataset of arbitration cases from Türkiye and evaluated across five different ML algorithms, with the Support Vector Machine (SVM) algorithm. Ye et al. (2023) investigated the causes of disputes in subcontracting practices by applying text mining and natural language processing (NLP) techniques to analyze 3150 publicly available litigation cases from China. Their methodology included the use of Term Frequency–Inverse Document Frequency (TF-IDF) and Singular Value

Decomposition (SVD) for data preprocessing, while K-Means++ was employed for clustering analysis, ultimately identifying 37 distinct causes of disputes.

12.3 Artificial Intelligence Technologies and Their Applications in Construction Dispute Management

12.3.1 Overview

A transformative revolution is in progress, driven by AI, which is reshaping social, economic, and political structures. In particular, AI is changing the way humans live and experience everyday life (Walsh 2017). AI is booming, and it is predicted that by 2050, AI through intelligent machines will be able to perform any intellectual task a human can perform. AI encompasses several subfields, with ML, DL, NLP, and LLMs being the most prominent (Santosh and Gaur 2021). ML is the scientific study of algorithms and statistical models that computer systems use to perform a specific task without being explicitly programmed (Mahesh 2020). In recent years, LLMs have emerged as a game-changing technology, displaying an extraordinary ability in NLP applications (Khan et al. 2023). These AI technologies rely on different algorithms and a large amount of data to solve problems from different industries, including the construction industry (Mathews 2019).

ML focuses on the task of enabling computational systems to learn from data about specific tasks automatically (Bilal et al. 2016). ML models can automatically learn and improve from data without being explicitly programmed (Alzubi 2023). There are different types of ML techniques, such as supervised, unsupervised, and reinforcement learning (Alzubi 2023). Supervised learning involves training a model to map inputs to outputs using labeled data, requiring external guidance. In contrast, unsupervised learning identifies patterns in unlabeled data without external input, commonly used for clustering and feature reduction by learning and applying patterns to new data. Reinforcement learning is an area of ML concerned with how software agents ought to take actions in an environment in order to maximize some notion of cumulative reward (Mathews 2019).

DL builds complex ideas from simpler ones using layered representations.
Each layer acts like a step in a computer program, processing information.
Deeper networks can perform more steps, learning more complex patterns.

Later layers use the results of earlier ones, making learning more powerful (Bengio et al. 2017). Most of the popular DL techniques are built on an Artificial Neural Network (ANN). DL has had a tremendous impact on various applications, such as computer vision, speech recognition, and NLP. ANNs are particularly well-suited for tasks involving classification and function estimation. An ANN functions as an information-processing system composed of numerous interconnected processing units known as neurons. These neurons are typically organized into multiple layers and are linked through weighted connections. The structure of an ANN includes an input layer, where data is fed into the network, and an output layer, which generates the network's response based on the input. Between these, one or more intermediate layers, referred to as hidden layers, play a crucial role in enabling the network to learn and represent complex relationships between data patterns. Each neuron in a layer is generally connected to every neuron in the subsequent layer, allowing the network to model intricate and nonlinear associations effectively (Yin et al. 2003). DL has revolutionized NLP by enabling models to learn language patterns directly from data. DL models, such as recurrent neural network (RNN), TextCNN, and transformers, are popular backbone architectures in NLP tasks.

The rapid development of AI has significantly impacted the field of NLP, which deals with analyzing and understanding text data (Pham and Han 2023). NLP is the subfield of both ML and linguistics that can process, understand, and simulate human language abilities through computational techniques. NLP can perform: (1) an array of tasks to tackle with text processing, including tokenization, stemming, and part of speech tagging (POS); (2) information retrieval (IR), information extraction (IE), and speech recognition; and (3) higher-level text analysis using both rule-based and ML methods (Shamshiri et al. 2024). NLP's core tasks are typically categorized into natural language understanding (NLU) and natural language generation (NLG). NLU involves encoding unstructured text into structured representations to help machines comprehend human language similarly to how humans do. In contrast, NLG focuses on converting structured data into coherent text, enabling machines to produce language that is easily understood by people. The main applications of NLG include machine translation (MT), answer generation in a Q&A system, and auto-summary, such as generating reports from company income data, summarizing the main content of an image, etc. (Ding et al. 2022). The major technologies and models of NLG include long short-term memory (LSTM), gated recurrent unit (GRU), Seq2seq, bidirectional encoder representations from transformers (BERT), and generative pre-trained transformer (GPT).

LLMs have garnered significant attention due to their ability to perform a wide range of NLP tasks with high accuracy. Notably, models such as ChatGPT, Gemini, Claude, and LLaMA exemplify the progress in this domain (Heo and Na 2025). LLMs are trained on massive amounts of text data from various sources, such as books, articles, websites, and social media. They can capture natural-language semantic and syntactic patterns and relationships (Taiwo et al. 2024). These models require the addition of customized layers or fine-tuning for specific tasks, as domain-specific language or knowledge is often lacking in the pre-trained model.

Training of LLMs typically utilizes unsupervised learning methods, where the model learns to predict the next word in a sequence, enabling it to produce fluent and context-aware text. Further fine-tuning on specialized datasets improves the model's effectiveness in specific applications, such as question answering, text generation, summarization, translation, and conversational systems (Heo and Na 2025). In the construction industry, there is a growing interest in the powerful capabilities of LLMs such as ChatGPT. These models are capable of answering a wide array of questions, making them extremely useful (Ghimire et al. 2023). Apart from question-and-answering, GPT models can assist with drafting project reports, generating safety checklists, summarizing contract clauses, regulations, and building codes, creating request-for-information templates, and translating technical documents across languages.

12.3.2 An AI Methodology to Develop Predictive Models

AI methodology typically involves several key steps: data collection, data preprocessing, feature engineering, model development, model evaluation, and model deployment, as depicted in Figure 12.1.

The first step is data collection. The quality of data is the key factor in determining the outputs of every data-driven model. Hence, data collection is a process to be carried out with utmost care and accuracy. Dataset preprocessing generally involves the following tasks: data cleaning, data integration, normalization, and data splitting. Raw data is usually characterized by several irregularities, such as missing values, noise, inconsistencies, and redundancies, whose presence influences the performance of subsequent learning steps. Then, the dataset is split into different sets: training, testing, and validation. The next step is feature engineering, which includes feature extraction and feature selection. Feature extraction reduces data dimensions to capture key aspects of high-dimensional data, while feature selection picks the most relevant features from the original set for model

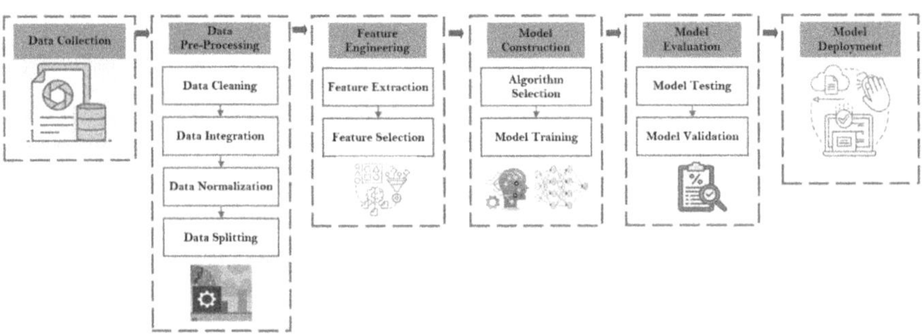

Figure 12.1 AI methodology. *Source:* Adapted from Perera et al. (2024).

use. In the next phase, suitable algorithm(s) will be selected based on the given dataset and the defined target objectives. AI algorithms automate the process of learning a model that identifies and captures the relationship between descriptive features and target variables within the data. Before deployment, the system must undergo rigorous testing and evaluation using data it has not encountered during training. This evaluation often involves comparing the model's performance against that of human experts, using carefully selected metrics to assess accuracy and effectiveness.

12.3.3 Construction Project Data

Construction projects involve a variety of data types, including structured, semi-structured, and unstructured data sources. The semi-structured or unstructured data text documents include contract documents, specifications, change orders, requests for information, schedules, cost estimates, meeting minutes, BOQ, and other correspondence; unstructured multimedia data such as 2D or 3D drawings, binary pictures, as well as audio and video files. These documents act as vital repositories of project knowledge. In addition, online discussions, emails, and progress reports are essential components of any information or knowledge management system.

Structured data refers to well-organized, finely processed information that is consistently categorized and easily stored in a tabular format of rows and columns within databases or spreadsheets (Mishra and Misra 2017). In the context of the construction industry, structured data typically includes basic project information, which includes project characteristics. Project characteristics include complexity of the project, time constraints, procurement method, type of client, size of the project, site factors, type of contract, source of funding, project cost, payment method, project scope, duration of the project, type of project, and project location.

The emergence of recent technologies enables all project-related electronic documents to be stored, updated, and shared through a common web server, thus enabling all parties involved to have access to project documents in real time (Guo et al. 2021). Document management systems significantly enhance document distribution efficiency, accelerate communication processes, and boost overall work productivity by minimizing time spent on paper-based document transmission while improving document quality. These systems primarily store and process documents digitally, encouraging a paperless work environment and minimizing physical storage requirements. Several document management systems predominantly used in construction projects are now available, including Aconex, ProjectWise 365, Autodesk BIM 360, Mi-Doc, Jira, Viewpoint, Speedbrick, Microsoft SharePoint, Proficient, Bluebeam, BuildSpace, and Trello. Thus, the construction industry is now flourishing with digital data and has thereby opened up new opportunities for implementing AI technologies in various applications.

12.4 AI Applications in the Construction Industry

AI, as a forefront digital innovation, has greatly enhanced business functions, service workflows, and overall industry efficiency across multiple sectors. By implementing AI techniques, organizations gain access to automated solutions and enjoy a competitive edge over traditional methods. Considering the success of AI in other industries like healthcare and manufacturing, there is immense potential to apply AI techniques in the construction industry. Several studies have explored the application of AI technologies in the construction industry, as outlined below.

12.4.1 Design and Planning

Perez and Tah (2023) introduced a pipeline that leverages AI and computer vision techniques to automate the measurement of as-built components. The system is designed to determine the dimensions of selected objects or components from a single image taken using a stereo camera setup. This approach provides an effective and accurate solution for measuring as-built components within constructed assets. Assaad et al. (2023) proposed a novel white-box expert knowledge-guided Bayesian Belief Network (BBN) with 90.51% accuracy to estimate bridge pile capacity. Wang, K et al. (2023) found that the random forest – levy-flight chaos sparrow search algorithm – back-propagation (RF-LCSSA-BP) model demonstrated superior predictive capabilities for the compressive strength of high-performance concrete. This model was found to outperform traditional methods, showcasing advantages in terms of both goodness of fit and prediction accuracy.

12.4.2 Construction Management

Salleh et al. (2023) identified a standard set of attributes for ML models in predicting building cost using systematic clarification. This is beneficial to improve the accuracy of estimation by providing the essence of developing a building cost estimation of ML. Zheng et al. (2023) developed a prediction index system and a prediction model based on the Random Forest (RF) algorithm. The predictive model holds the potential to significantly enhance the accuracy of construction cost predictions, thereby offering a foundation for optimizing the overall cost management of construction projects. Ali and Burhan (2023) found that the XGBoost algorithm exhibited superior performance, boasting a 92% accuracy rate in selecting the most correlated variables for project cost. This suggests that the XGBoost model can be a valuable asset for project managers, enabling the identification of influential factors and the accurate estimation of building project costs. Gondia et al. (2020) concluded that a naïve Bayesian model provides a better predictive performance than a decision tree in predicting delays in construction projects using objective data sources, such as relevant delay sources and factors, a multivariate dataset of previous projects' time performance, and delay-inducing risk sources.

12.4.3 Safety and Risk Management

The construction industry remains one of the most hazardous sectors, with accidents occurring at varying levels of severity. Koc et al. (2022) integrated the discrete wavelet transform with various ML algorithms to predict the occurrence of occupational accidents. The time series dataset used for this study included 393,160 documented occupational accidents in Turkey, covering the period from 2012 to 2020. Mammadov et al. (2023) developed a set of predictive ML models as part of a proactive forecasting strategy within a prescriptive framework designed to prevent construction accidents or reduce their severity. The study highlighted low awareness of workplace hazards and improper operation of vehicles as the most significant immediate causes. At the same time, inadequate risk recognition and insufficient site supervision were identified as the main root causes of accidents, particularly in pipeline construction projects. Tixier et al. (2016) found RF and Stochastic Gradient Tree Boosting (SGTB) could produce accurate forecasts to predict the likely outcomes of an accident in terms of injury type, energy type, and body part. Liu et al. (2023) proposed a model designed to predict workers' fall risk by classifying their state into three categories: steady, unsteady, and fallen. The model also provides corresponding confidence probabilities for each classification, offering a quantitative measure of prediction certainty.

12.4.4 Quality Control and Maintenance

Chen et al. (2024) claimed that traditional manual compliance checking methods are inefficient and error-prone, failing to meet modern engineering demands. Therefore, the authors introduce an innovative automated compliance-checking framework that integrates LLMs, DL models, and ontology knowledge models. Zhang and El-Gohary (2021) proposed a deep neural network-based approach for advanced IE, aiming to retrieve both semantic and syntactic elements from regulatory documents within the AEC domain. The effectiveness of the method was assessed through its application to various regulatory documents relevant to the AEC industry. Mariniuc et al. (2024) introduced an innovative method for semi-automated building inspection, aimed at reducing the time required to evaluate maintenance needs. This approach combines ML techniques with 3D scanning technologies to enhance the efficiency and accuracy of building assessments.

12.4.5 Sustainability and Energy Efficiency

A predictive model for green building design was developed using ML and DL techniques to optimize resource usage, enhance occupant comfort, and minimize the environmental impact throughout the green building design process (Mahmood et al. 2024). Alijoyo (2024) proposed an AI-driven DL framework employing convolutional neural networks (CNNs) to improve energy management in smart buildings. This framework utilizes DL to analyze historical energy consumption data, weather patterns, and building-specific characteristics to accurately forecast future energy usage. It also incorporates anomaly detection mechanisms to identify inefficiencies and faults within the energy management system. By enabling real-time energy optimization and supporting demand response strategies, the framework aims to lower energy costs and boost the overall operational efficiency of smart buildings. Xiang et al. (2022) developed an AI-based energy management model (AI-EMM) for green buildings, designed to benefit both commercial and residential structures. The AI-EMM is being developed as a plug-and-play device to enable easy installation and integration. Its automation is driven by multiuser preferences, incorporating smart user identification and real-time environmental observations as input. This approach aims to enhance user comfort while maximizing energy efficiency.

Alongside, AI technologies and their applications have widened to predict and mitigate disputes in the construction industry in recent years. The next section provides a detailed explanation of the AI application in managing disputes proactively.

12.5 AI Applications in Dispute Management

Integrating AI into proactive dispute management stands out as a timely development, and over the last few decades, this has attracted researchers as well. This section reviews the existing studies, especially journal articles from the last decade on dispute management incorporating AI and its various technologies. Several studies (19) have been identified and summarized in Table 12.2.

Table 12.2 summarizes the studies on AI applications in dispute management in terms of AI technologies adopted, prediction outcome, algorithms utilized, and the number of studies. The majority of the studies utilized ML technology compared to DL and NLP. Many studies predicted various outcomes, which include dispute outcome (in favor of which party); dispute resolution methods; compensation of disputes: No., cost, time, both cost and time; dispute Index; dispute or no dispute; litigation outcome, and claim allowed, rejected, or partially allowed. Zheng et al. (2021) investigated the primary causes of disputes in public-private partnership (PPP) projects, analyzed the outcomes of these disputes, and developed a predictive model for litigation results using data from 171 PPP litigation cases

Table 12.2 Summary of AI applications in dispute management.

AI technology	Prediction outcome	Algorithms utilized	Number of studies	Sources
Machine learning	Dispute outcome (in favor of which party) Dispute resolution methods Compensation of disputes: No., cost, time, both cost and time Dispute Index Dispute or no dispute, Ligation outcome, Claim allowed/ rejected/partially allowed	Logistic Regression (LR), Support Vector Machines (SVM), Polynomial Kernel SVM, Radial Basis Function (RBF) SVM, Decision Trees (DT), Classification and Regression Tree (CART), C4.5, C5.0, Quick, Unbiased, and Efficient Statistical Tree (QUEST), Chi-Square Automatic Interaction Detector (CHAID), K-Nearest Neighbors (K-NN), Naive Bayes (NB), Random Forest (RF), Gradient Boosting Decision Tree (GBDT), Bayesian Belief Network (BBN), Tree-Augmented Naïve Bayesian (TAN), Multilayer Perceptron (MLP), SVM and Artificial Neural Network (ANN) Hybrid, SVM + C5.0 Hybrid, Fast Messy Genetic Algorithm (fmGA), and SVM + fmGA + Fuzzy Logic (FL)	11	c, f, h, g, n, i, o, p, q, r, s
Machine learning and deep learning	Litigation outcome	DT, Random Forest (RF), SVM, Neural Network (NN), Artificial Neural Networks (ANN)	2	d, m
Deep learning	Causes of disputes	ANN	1	j
Natural language processing	Categories of legal cases, Litigation outcome, Causes of disputes, 14 categories (topics) of defect cases, Category of legal factors	Latent Dirichlet allocation, Multilabel Text Classification Models, SVM, LR, NB, DT, K-NN, TextCNN, TextCNN + LSTM, RCNN, Transformer, Term Frequency-Inverse Document Frequency (TF-IDF), Singular Value Decomposition (SVD), K-Means++, Term Frequency, Logarithmic Term Frequency, Augmented Term Frequency, Projective Adaptive Resonance Theory (PART)	5	a, b, e, k, l

a: Elelu et al. (2024), b: Zhong et al. (2024), c: Un et al. (2024), d: Alqaisi et al. (2024), e: Ye et al. (2023), f: Ayhan et al. (2023), g: Wang et al. (2022), h: Ayhan et al. (2022), i: Zheng et al. (2021), j: Fatima et al. (2019), k: Jallan et al. (2019), l: Mahfouz et al. (2018), m: Anysz et al. (2021), n: Ayhan et al. (2021), o: Chaphalkar et al. (2015), p: Chou et al. (2014), q: Chou et al. (2013b), r: Chou et al. (2013a), s: Chou and Lin (2013) Ayhan et al. (2021), o: Chaphalkar et al. (2015), p: Chou et al. (2014), q: Chou et al. (2013b), r: Chou et al. (2013b), s: Chou and Lin (2013).

sourced from China Judgments. The study found that an ensemble model combining Gradient Boosting Decision Tree (GBDT), K-Nearest Neighbors (K-NN), and a Multilayer Perceptron (MLP) neural network outperformed 9 other individual ML models, achieving a high prediction accuracy of 96.42%. Wang et al. (2022) focused on the prevention of construction disputes by developing paired mechanistic and empirical models to evaluate the feasibility of achieving a mechanistic understanding of such conflicts. The study introduced a predictive model to estimate dispute propensity using a dispute index scale: 0–3 indicating early warning, 4–10 as a warning, and values above 10 as an emergency. Under ideal conditions, the model achieved an accuracy rate exceeding 95%.

Few studies utilized ML and DL models. Alqaisi et al. (2024) developed a predictive model to forecast the outcomes of construction change order disputes using historical case data retrieved through keyword searches from online legal databases such as Westlaw and Lexis Nexis. This model offers valuable benefits to the construction industry by enabling early identification of potential risks, allowing for timely interventions to prevent disputes from escalating. Furthermore, the model can assist involved parties in negotiating settlements before resorting to litigation, thereby minimizing legal expenses and reducing delays. Anysz et al. (2021) developed an ML model using decision trees and ANNs to predict the result of a dispute. Thus, the model assists in ascertaining the correctness of the decision to choose litigation or not to resolve a dispute between the general contractor and the investor. Another study utilized a DL model to develop the dispute attribute prediction model based on ANNs (Fatima et al. 2019).

There seems to be a growing trend in the use of NLP, which extracts information from the text-based data available in construction projects and makes predictions supporting the decision-making of stakeholders. As per Table 12.2, the prediction models were developed to predict categories of legal cases, litigation outcomes, causes of disputes, categories of defect cases, and categories of legal factors using data such as utility manuals, judgment documents, litigation cases, and case summaries. In particular, accessible case documents play a vital role in developing predictive models using NLP. For example, Zhong et al. (2024) developed models for automatic legal text classification for construction contract dispute cases using five SL-based models (i.e., SVM, LR, NB, DT, and K-NN) and four DL-based models (i.e., TextCNN, TextCNN + LSTM, RCNN, and transformer). Experimental results on legal text classification show that the TextCNN classifier produced the best performance among others. Ye et al. (2023) investigated the causes of disputes in subcontracting practices by automatically examining 3150 litigation cases that are publicly available in China using text mining and NLP. Thus, the review shows the potential and the trend of AI technologies in managing disputes in the construction industry. The next section of the chapter develops an AI-driven framework for proactive dispute management in the construction industry.

12.6 An AI-Driven Framework for Proactive Dispute Management in the Construction Industry

Figure 12.2 depicts the AI-driven framework for proactive dispute management in the construction industry. The center of the framework is "dispute." The typology of disputes lays a foundation for understanding the disputes and is also important for AI model development. For any AI model development, data is the key. The construction project data can be categorized in two major ways: structured/semi-structured and unstructured data. The unstructured data consists of the text-based project correspondence flows between the construction stakeholders. The unstructured data can be litigation cases, judgment summaries, contract clauses, contract documents, variation orders, requests for information, meeting minutes, site diaries, notices, and other correspondences that can be used for model development. The structured and semi-structured data, which are categorical in nature, include project attributes, such as project characteristics (e.g., project type, project cost, duration, procurement method, type of client, project size, site factors, type of contract, source of funding, etc.) and

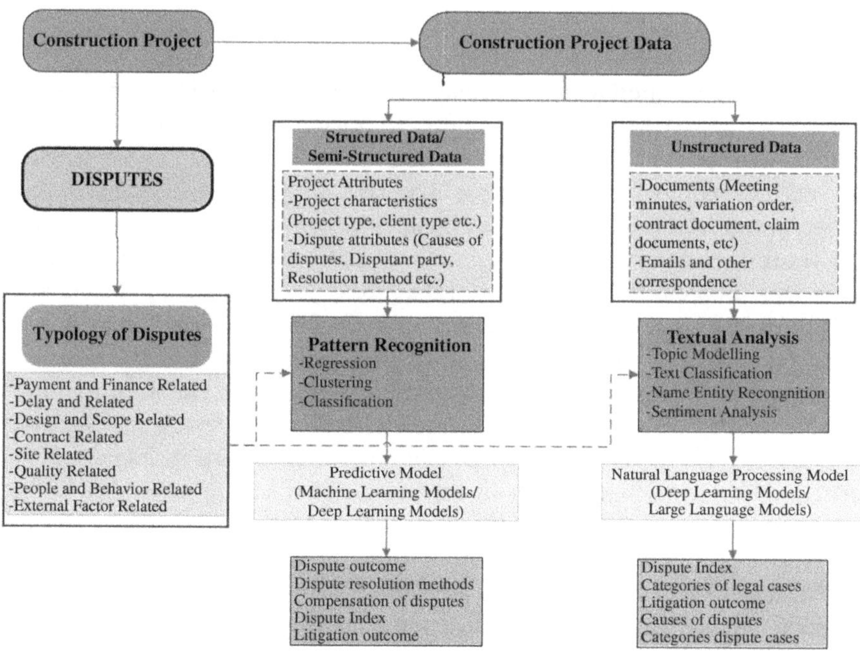

Figure 12.2 AI-driven framework for proactive dispute management in the construction industry.

dispute-related aspects, including the causes of disputes, resolution methods, and disputant parties. As depicted, NLP concepts, such as topic modeling, named entity recognition, text classification, and sentiment analysis, can be employed to analyze textual data. NLP technology will be utilized to train the unstructured data, whereas traditional predictive modeling will be used for the categorical data. Under NLP technology, DL and LLMs will be used to train the textual data, while ML and DL algorithms will be used to train the structured data. The typology of disputes plays a crucial role in labeling data for the development of AI models, particularly when using supervised learning approaches. Accurate categorization of dispute types ensures that the model can be effectively trained to recognize patterns and make reliable predictions based on labeled historical data. Thus, the framework provides a foundation for proactive dispute management in the construction industry using AI technologies and concepts.

12.7 Conclusion

Disputes are a frequent and well-recognized occurrence in the construction industry. Far from being immune, the industry has, in fact, seen its growth influenced by the presence of such conflicts. Known for its highly litigious environment, the construction sector often experiences disputes among various stakeholders, which has almost become routine. These disputes can negatively affect both the cost and timeline of construction projects. To address this issue, industry professionals and academic researchers have explored various methods to either prevent disputes from occurring or to resolve them once they arise. A number of dispute avoidance strategies and established resolution techniques are widely implemented in practice. However, recent studies highlight that disputes continue to pose significant challenges, with effects that go beyond the project level, impacting stakeholders and potentially influencing the broader national economy. In recent years, several initiatives have been carried out by industry practitioners and researchers to combine digital technologies with existing dispute management practices. AI shows the potential for predicting disputes before they occur and thereby managing disputes effectively. Therefore, the chapter developed an AI-driven framework for proactive dispute management in the construction industry.

This chapter first reviews the disputes and related issues and develops the theoretical layout for the typology of construction disputes. Then, the chapter investigated AI technologies, AI methodology, and AI applications in the construction industry with a key focus on dispute management. The structured data is used to train the ML and DL models, while the unstructured textual data is used in NLP models, especially DL and LLMs. The developed framework incorporates the typology of causes of disputes, construction project data, AI concepts, AI technologies like ML, DL, and NLP, LLMs, and potential outcomes of the AI models. This framework can be utilized by future researchers to develop dispute predictive models aiming to predict the likelihood of disputes, dispute resolution methods, causes of disputes, and other dispute-related attributes. Thus, using AI for dispute prediction could offer a proactive approach to both avoiding and effectively managing disputes in the construction industry.

References and Suggested Further Reading

Abdul-Malak, M.-A.U. and Senan, M.H. (2020). Operational mechanisms and effectiveness of adjudication as a key step in construction dispute resolution. *Journal of Legal Affairs and Dispute Resolution in Engineering and Construction* 12 (1): 04519051.

Ahmed, M.O. and El-adaway, I.H. (2023). Data-driven analysis of construction bidding stage–related causes of disputes. *Journal of Management in Engineering* 39 (5): 04023026.

Ali, Z.H. and Burhan, A.M. (2023). Hybrid machine learning approach for construction cost estimation: an evaluation of extreme gradient boosting model. *Asian Journal of Civil Engineering* 1–16.

Alijoyo, F.A. (2024). AI-powered deep learning for sustainable industry 4.0 and internet of things: Enhancing energy management in smart buildings. *Alexandria Engineering Journal* 104: 409–422.

Alqaisi, A.S., Ataei, H., Seyrfar, A., and Al Omari, M. (2024). Predicting the outcome of construction change disputes using machine-learning algorithms. *Journal of Legal Affairs and Dispute Resolution in Engineering and Construction* 16 (1): 04523051.

Alrasheed, K.A., Soliman, E., and AlMesbah, F.E. (2023). Dispute classification in construction projects based on litigation cases. *Journal of Legal Affairs and Dispute Resolution in Engineering and Construction* 15 (3): https://doi.org/10.1061/JLADAH.LADR-910.

Alzubi, Y. (2023). Comparison of various machine learning models for estimating construction projects sales valuation using economic variables and indices. *Journal of Soft Computing in Civil Engineering* 8 (1): 1–32.

Anysz, H., Apollo, M., and Grzyl, B. (2021). Quantitative risk assessment in construction disputes based on machine learning tools. *Symmetry* 13 (5): 744.

Assaad, R.H., Hu, X., and Hussein, M. (2023). Expert knowledge–guided Bayesian belief networks for predicting bridge pile capacity. *Journal of Bridge Engineering* 28 (9): 04023058.

Ayhan, M., Dikmen, I., and Birgonul, M.T. (2021). Predicting the occurrence of construction disputes using machine learning techniques. *Journal of Construction Engineering and Management* 147 (4): 04021022.

Ayhan, M., Dikmen, I., and Birgonul, M.T. (2022). Comparing performances of machine learning techniques to forecast dispute resolutions. *Teknik Dergi* 33 (5): 12577–12600.

Ayhan, M., Dikmen, I., and Birgonul, M. T. (2023). Disputes Using Machine Learning Techniques Classifying Compensations in Construction.

Barnett, J. and Treleaven, P. (2018). Algorithmic dispute resolution—the automation of professional dispute resolution using AI and blockchain technologies. *Computer Journal* 61 (3): 399–408. https://doi.org/10.1093/comjnl/bxx103.

Bengio, Y., Goodfellow, I., and Courville, A. (2017). *Deep Learning*, vol. 1. Cambridge, MA: MIT Press.

Bilal, M., Oyedele, L.O., Qadir, J. et al. (2016). Big Data in the construction industry: a review of present status, opportunities, and future trends. *Advanced Engineering Informatics* 30 (3): 500–521.

Cappiello, B. and Carullo, G. (2021). *Blockchain, Law and Governance.* Cham: Springer.

Chaphalkar, N., Iyer, K., and Patil, S.K. (2015). Prediction of outcome of construction dispute claims using multilayer perceptron neural network model. *International Journal of Project Management* 33 (8): 1827–1835.

Chen, N., Lin, X., Jiang, H., and An, Y. (2024). Automated building information modeling compliance check through a large language model combined with deep learning and ontology. *Buildings* 14 (7): 1983.

Chou, J.-S. and Lin, C. (2013). Predicting disputes in public-private partnership projects: Classification and ensemble models. *Journal of computing in civil engineering* 27 (1): 51–60.

Chou, J.-S., Cheng, M.-Y., and Wu, Y.-W. (2013a). Improving classification accuracy of project dispute resolution using hybrid artificial intelligence and support vector machine models. *Expert Systems with Applications* 40 (6): 2263–2274.

Chou, J.-S., Tsai, C.-F., and Lu, Y.-H. (2013b). Project dispute prediction by hybrid machine learning techniques. *Journal of civil engineering and management* 19 (4): 505–517.

Chou, J.-S., Cheng, M.-Y., Wu, Y.-W., and Pham, A.-D. (2014). Optimizing parameters of support vector machine using fast messy genetic algorithm for dispute classification. *Expert Systems with Applications* 41 (8): 3955–3964.

Ding, Y., Ma, J., and Luo, X. (2022). Applications of natural language processing in construction. *Automation in Construction* 136: 104169.

Elelu, K., Do, Q., Le, T., and Piratla, K. (2024). Sample of legal provision retrieval using topic clustering to support utility relocation agreement drafting in highway projects. *Journal of Legal Affairs and Dispute Resolution in Engineering and Construction* 16 (2): 04523058.

El-Sayegh, Ahmad, I., Aljanabi, M. et al. (2020). Construction disputes in the UAE: causes and resolution methods. *Buildings (Basel)* 10 (10): 171–115. https://doi.org/10.3390/buildings10100171.

Faraji, A., Rashidi, M., Tezangi, M.R., and Perera, S. (2021). Multihybrid dispute resolution framework for projects of downstream sector of petroleum industry. *Journal of Legal Affairs and Dispute Resolution in Engineering and Construction* 13 (4): 04521026.

Fatima, A., Prasad, B.S.K., and Sekhar, T.S. (2019). Prediction of construction dispute using artificial neural networks testimonies from Indian construction projects. *International Journal of Civil Engineering and Technology* 10 (1): 582–594.

Francis, M., Ramachandra, T., and Perera, S. (2022). Disputes in construction projects: a perspective of project characteristics. *Journal of Legal Affairs and Dispute Resolution in Engineering and Construction* 14 (2): https://doi.org/10.1061/(ASCE)LA.1943-4170.0000535.

Francis, M., Perera, S., Zhou, W., and Nanayakkara, S. (2025). A comprehensive typology of causes leading to disputes in construction projects. *Journal of Legal Affairs and Dispute Resolution in Engineering and Construction* 17 (4): 03125006.

Ghimire, P., Kim, K., and Acharya, M. (2023). Generative AI in the construction industry: opportunities & challenges. *arXiv preprint arXiv:2310.04427*.

Gondia, A., Siam, A., El-Dakhakhni, W., and Nassar, A.H. (2020). Machine learning algorithms for construction projects delay risk prediction. *Journal of Construction Engineering and Management* 146 (1): 04019085.

Gregory-Stevens, J., Frame, I., and Henjewele, C. (2016). Mediation in construction disputes in England. *International Journal of Law in the Built Environment* 8 (2): 123–136.

Guo, F., Jahren, C.T., and Turkan, Y. (2021). Electronic document management systems for the transportation construction industry. *International Journal of Construction Education and Research* 17 (1): 52–67.

Heo, S. and Na, S. (2025). Ready for departure: Factors to adopt large language model (LLM)-based artificial intelligence (AI) technology in the architecture, engineering and construction (AEC) industry. *Results in Engineering* 25: 104325.

Jagannathan, M. and Delhi, V.S.K. (2020). Litigation in construction contracts: literature review. *Journal of Legal Affairs and Dispute Resolution in Engineering and Construction* 12 (1): 03119001.

Jallan, Y., Brogan, E., Ashuri, B., and Clevenger, C.M. (2019). Application of natural language processing and text mining to identify patterns in construction-defect litigation cases. *Journal of Legal Affairs and Dispute Resolution in Engineering and Construction* 11 (4): 04519024.

Kalogeraki, M. and Antoniou, F. (2024). Claim management and dispute resolution in the construction industry: current research trends using novel technologies. *Buildings* 14 (4): 967.

Khan, N., Kimito, E. C., Tran, S. et al. (2023). Chapter Extracting Information from Construction Safety Requirements Using Large Language Model.

Khan, K.K.A., Siyal, A.W., and Hussain, S. (2025). Dispute resolution management in construction projects: a stakeholders centric investigation. *Journal of Legal Affairs and Dispute Resolution in Engineering and Construction* 17 (3): 04525020.

Koc, K., Ekmekcioğlu, Ö., and Gurgun, A.P. (2022). Accident prediction in construction using hybrid wavelet-machine learning. *Automation in Construction* 133: 103987.

Kukah, A.S.K., Jin, X., Osei-Kyei, R., and Perera, S. (2025). Legal dispute resolution techniques for carbon trading in the construction industry: a comprehensive review and future directions. *Journal of Legal Affairs and Dispute Resolution in Engineering and Construction* 17 (3): 04525033.

Liu, X., Xu, F., Zhang, Z., and Sun, K. (2023). Fall-portent detection for construction sites based on computer vision and machine learning. *Engineering, Construction and Architectural Management* 32 (3): 1499–1521.

Mahesh, B. (2020). Machine learning algorithms – a review. *International Journal of Science and Research (IJSR) [Internet]* 9 (1): 381–386.

Mahfouz, T., Kandil, A., and Davlyatov, S. (2018). Identification of latent legal knowledge in differing site condition (DSC) litigations. *Automation in Construction* 94: 104–111.

Mahmood, S., Sun, H., El-Kenawy, E.-S.M. et al. (2024). Integrating machine and deep learning technologies in green buildings for enhanced energy efficiency and environmental sustainability. *Scientific Reports* 14 (1): 20331.

Mammadov, A., Kazar, G., Koc, K., and Tokdemir, O.B. (2023). Predicting accident outcomes in cross-border pipeline construction projects using machine learning algorithms. *Arabian Journal for Science and Engineering* 48: 13771–13789.

Mariniuc, A.M., Cojocaru, D., and Abagiu, M.M. (2024). Building surface defect detection using machine learning and 3D scanning techniques in the construction domain. *Buildings* 14 (3): 669.

Mathews, S.M. (2019). Explainable artificial intelligence applications in NLP, biomedical, and malware classification: a literature review. In: *Intelligent Computing* (ed. K. Arai, R. Bhatia, and S. Kapoor), 1269–1292. Springer.

Mishra, S. and Misra, A. (2017). Structured and unstructured big data analytics. *2017 International Conference on Current Trends in Computer, Electrical, Electronics and Communication (CTCEEC)*, Mysore, India (8–9 September 2017), 740–746. IEEE.

Moza, A. and Paul, V.K. (2017). Review of the effectiveness of arbitration. *Journal of Legal Affairs and Dispute Resolution in Engineering and Construction* 9 (1): 03716002.

Naji, K.K., Mansour, M.M., and Gunduz, M. (2020). Methods for modeling and evaluating construction disputes: a critical review. *IEEE Access* 8: 45641–45652.

Olatunji, O.A. (2016). Constructing dispute scenarios in building information modeling. *Journal of Legal Affairs and Dispute Resolution in Engineering and Construction* 8 (1): C4515001.

Perera, S., Francis, M., and Nanayakkara, S. (2024). Predicting disputes in the digital age: a machine learning approach to construction projects. *Built Environment Economist: Australia and New Zealand* 28–31.

Perez, H. and Tah, J.H. (2023). Towards automated measurement of as-built components using computer vision. *Sensors* 23 (16): 7110. https://mdpi-res.com/d_attachment/sensors/sensors-23-07110/article_deploy/sensors-23-07110.pdf?version=1691743243.

Pham, H.T. and Han, S. (2023). Natural language processing with multitask classification for semantic prediction of risk-handling actions in construction contracts. *Journal of Computing in Civil Engineering* 37 (6): 04023027.

Salleh, H., Wang, R., Affandi, N.Z.H., and Abdul-Samad, Z. (2023). Selecting a standard set of attributes for the development of machine learning models of building project cost estimation. *Planning Malaysia* 21: https://doi.org/10.21837/pm.v21i29.1359.

Santosh, K.C. and Gaur, L. (2021). Introduction to AI in public health. In: *Artificial Intelligence and Machine Learning in Public Healthcare* (ed. K.C. Santosh and L. Gaur), 1–10. Singapore: Springer.

Saseendran, A., Bigelow, B.F., Rybkowski, Z.K., and Jourdan, D.E. (2020). Disputes in construction: evaluation of contractual effects of consensusDOCS. *Journal of Legal Affairs and Dispute Resolution in Engineering and Construction* 12 (2): https://doi.org/10.1061/(ASCE)LA.1943-4170.0000377.

Saygili, M., Mert, I.E., and Tokdemir, O.B. (2022). A decentralized structure to reduce and resolve construction disputes in a hybrid blockchain network. *Automation in Construction* 134: 104056. https://doi.org/10.1016/j.autcon.2021.104056.

Senaratne, S. and Farhan, S. (2023). Role of standard contracts in mitigating disputes in construction. *Journal of Legal Affairs and Dispute Resolution in Engineering and Construction* 15 (1): 04522045.

Shamshiri, A., Ryu, K.R., and Park, J.Y. (2024). Text mining and natural language processing in construction. *Automation in Construction* 158: 105200.

Taiwo, R., Bello, I.T., Abdulai, S.F. et al. (2024). Generative AI in the construction industry: a state-of-the-art analysis. *arXiv preprint arXiv:2402.09939*.

Tao, X., Das, M., Liu, Y., and Cheng, J.C. (2021). Distributed common data environment using blockchain and interplanetary file system for secure BIM-based collaborative design. *Automation in Construction* 130: 103851.

Tixier, A.J.-P., Hallowell, M.R., Rajagopalan, B., and Bowman, D. (2016). Application of machine learning to construction injury prediction. *Automation in Construction* 69: 102–114.

Torkanfar, N., Rezazadeh Azar, E., and McCabe, B. (2023). A blockchain-based approach for managing construction claims. *European Council on Computing in Construction* 4: 1–11.

Un, B., Erdis, E., Aydınlı, S. et al. (2024). Forecasting the outcomes of construction contract disputes using machine learning techniques. *Engineering, Construction and Architectural Management*. https://doi.org/10.1108/ECAM-05-2023-0510.

Walsh, T. (2017. The AI revolution. NSW Department of Education Education: Future Frontiers.

Wang, P., Huang, Y., Zhu, J., and Shan, M. (2022). Construction dispute potentials: mechanism versus empiricism in artificial neural networks. *Sustainability* 14 (22): 15239.

Wang, J., Zhang, S., Fenn, P. et al. (2023). Adopting BIM to facilitate dispute management in the construction industry: a conceptual framework development. *Journal of Construction Engineering and Management* 149 (1): 03122010.

Wang, K., Ren, J., Yan, J. et al. (2023). Research on a concrete compressive strength prediction method based on the random forest and LCSSA-improved BP neural network. *Journal of Building Engineering* 76: 107150.

Xiang, Y., Chen, Y., Xu, J., and Chen, Z. (2022). Research on sustainability evaluation of green building engineering based on artificial intelligence and energy consumption. *Energy Reports* 8: 11378–11391.

Ye, Y.-X., Shan, M., Gao, X. et al. (2023). Examining causes of disputes in subcontracting litigation cases using text mining and natural language processing techniques. *International Journal of Construction Management* 1–13.

Yin, C., Rosendahl, L., and Luo, Z. (2003). Methods to improve prediction performance of ANN models. *Simulation Modelling Practice and Theory* 11 (3–4): 211–222.

Zand Pazandi, A., Forootan, F., Pourrostam, T., and Ravanshadnia, M. (2024). Arbitration of disputes in the construction industry. *Journal of Legal Affairs and Dispute Resolution in Engineering and Construction* 16 (4): 04524021.

Zhang, R. and El-Gohary, N. (2021). A deep neural network-based method for deep information extraction using transfer learning strategies to support automated compliance checking. *Automation in Construction* 132: 103834.

Zheng, X., Liu, Y., Jiang, J. et al. (2021). Predicting the litigation outcome of PPP project disputes between public authority and private partner using an ensemble model. *Journal of Business Economics and Management* 22 (2): 320–345.

Zheng, Z., Zhou, L., Wu, H., and Zhou, L. (2023). Construction cost prediction system based on random forest optimized by the Bird Swarm Algorithm. *Mathematical Biosciences and Engineering* 20 (8): 15044–15074. https://www.aimspress.com/article/10.3934/mbe.2023674.

Zhong, B., Shen, L., Pan, X. et al. (2024). 'Dispute classification and analysis: deep learning–based text mining for construction contract management. *Journal of Construction Engineering and Management* 150 (1): 04023151.

Chapter Review Questions

1) Which of the following is not an alternative dispute resolution (ADR) method commonly used in the construction industry?
   a) Mediation
   b) Arbitration
   c) Litigation
   d) Adjudication
2) Which of the following is the primary goal of mediation?
   a) Give a legally binding judgment
   b) Encourage parties to reach a mutually acceptable agreement
   c) Provide punishment to the guilty
   d) None of the above
3) _______________ ADR method provides a binding decision and involves a third-party expert in decision-making.
   a) Arbitration
   b) Negotiation
   c) Conciliation
   d) Mediation
4) The key advantage of mediation compared to arbitration is that:
   a) It guarantees a binding decision
   b) It is faster and focuses on maintaining relationships
   c) It requires a legal representative
   d) It is always governed by company policies
5) Artificial intelligence refers to;
   a) A technology that makes machines do things as humans
   b) A system that performs only mechanical tasks
   c) A data storage tool
   d) A program to perform numerical tasks
6) Which AI subfield especially understands and generates human language?
   a) Computer vision
   b) Natural language processing

   c) Machine learning
   d) Expert system
7) Computer vision is mainly used to:
   a) Interpret and analyze visual information from images
   b) Generate text
   c) Speech recognition
   d) Predict continuous numbers
8) Which of the following best describes deep learning?
   a) Multilayered neural networks to identify complex patterns
   b) Depends on shallow learning
   c) Requires rule-based systems
   d) Deep learning models can be developed without training
9) _______________ refers as autonomous machines that perform physical tasks.
   a) Computer vision
   b) Natural language processing
   c) Machine learning
   d) Robotics
10) Natural language processing allows computers to learn and understand human language, including speech and text. True/False
11) What is a large language model primarily designed to do?
   a) Generate and understand human-like text using models trained on a large corpus of data.
   b) Generate real-time images
   c) Understand human behavior and make predictions
   d) Produce expert systems
12) Which of the following is a core technique used in training LLMs like GPT?
   a) Encoder mechanism
   b) Reinforcement learning from human feedback
   c) Supervised learning
   d) Data encryption
13) Which of the following is NOT one of the eight major categories of construction disputes identified in the chapter?
   a) Payment and finance
   b) Delay and time
   c) Contract
   d) Marketing and business
14) Payment delays by the client will lead to:
   a) Payment- and finance-related disputes
   b) Delay- and time-related disputes
   c) Contract-related disputes
   d) Quality-related disputes

15) Frequent scope changes lead to:
    a) Payment- and finance-related disputes
    b) Delay- and time-related disputes
    c) Contract-related disputes
    d) Design- and scope-related disputes
16) Disputes arising due to delayed payments, cost overruns, or funding issues are classified as:
    a) Delay- and time-related disputes
    b) Payment- and finance-related disputes
    c) People- and behavior-related disputes
    d) Quality-related disputes
17) Poor communication, lack of teamwork, and opportunistic behavior of stakeholders commonly lead to:
    a) Delay- and time-related disputes
    b) Payment- and finance-related disputes
    c) People- and behavior-related disputes
    d) Quality-related disputes
18) Weather conditions, political instability, and changes in government regulations are examples of:
    a) Delay- and time-related disputes
    b) Payment- and finance-related disputes
    c) People- and behavior-related disputes
    d) External factors-related disputes
19) Which of the following is an example of unstructured project data used in dispute prediction models?
    a) Project duration
    b) Project cost
    c) Project correspondence, i.e., emails
    d) Site location details
20) In the AI-driven dispute prediction model, unstructured textual data is primarily used to:
    a) Train the deep learning models to predict disputes
    b) Generate site images
    c) Provide voice command to stakeholders
    d) Generate structural drawings

Exercises

1) Identify the ADR methods widely used in the construction industry, and list the advantages and disadvantages of each method.

2) List and briefly describe the major categories of disputes in the construction industry. For each category, provide one real-world example of a situation that could lead to such a dispute.
3) Identify the subfields of AI you are familiar with, and provide real-world examples illustrating the application of each technology.

Discussion

1) Among the eight major dispute classifications identified in the chapter, which do you consider the most critical type of dispute? Discuss the proactive measures that can be adopted to manage this type of dispute effectively.
2) "The availability of digital construction data has paved the way for artificial intelligence applications in the construction industry." Elaborate on this statement using relevant examples and supporting evidence.
3) How might the introduction of an AI-driven proactive dispute management framework assist construction professionals in managing disputes? Illustrate your discussion with a real-world construction project scenario.

13

Designating and Engaging Lead Proactive Dispute Manager(s) (PDM(s))

Readers' Learning Objectives and Outcomes

1) *Comprehend why designating one or two outstanding individuals as lead proactive dispute managers is critical for implementing PDM principles effectively within a project team.*
2) *Gain insights into methods for motivating and guiding project team members to anticipate potential disputes early and apply proactive measures to prevent, avoid, or douse them.*
3) *Acquire practical knowledge on the criteria for selecting lead PDMs, their responsibilities, and how to empower them to influence dispute management culture across the project lifecycle.*

13.1 Introduction

Proactive dispute management as a new discipline and doctrine calls for time and application experience to completely understand how best to drive it. Nonetheless, all project team members are invited and encouraged to participate in driving and entrenching its practice in the process. While a ship may have more than one captain, there will always be one master captain responsible for the entire ship in the singular sense of ultimate command to avoid dangerous confusion. This scenario is also applicable to PDM, where there may be two lead proactive dispute managers, but one must be the master captain with whom ultimate decisions reside under critical or crucial situations.

In designating and engaging lead PDMs, the perils of the briefly successful "partnering" workshops and deployment must be remembered and avoided. Partnering was once so successful that the now-defunct Dispute Avoidance and Resolution Task Force (DART) in the construction industry was created under a pilot program with the American Arbitration Association. According to Daniel Yamshon, a partnering consultant and workshops convener over three decades ago, "Its success was its downfall, since each [partnering] workshop had to be designed for the specific project, the design process requiring interviews with all the stakeholders, the owner, the design team, the general contractor, the subs [subcontractors] and sometimes the material suppliers, it was relatively expensive although demonstrably cost-effective." In addition, the regimented approach to partnering and involvement of external experts detached from individual projects promoted the erroneous one-size-fits-all in construction and related industries where such doesn't apply. As a result, partnering was viewed as an extra significant cost that didn't match or justify its ballpark value in the process, and it naturally got written off as a fad and had to die and become largely abandoned.

In contrast, PDM recognizes the uniqueness of every project and does not treat any with a one-size-fits-all approach. It doesn't involve external people in the project, and it focuses away from resource-consuming workshops. The immediate project actors are the PDM implementation drivers. Hence, every employee and every member of any project team should be made aware and knowledgeable about PDM from the onset by sharing extant PDM processes and procedures and the grounds upon which they are anchored, such as organizational policies and procedures; internal quality assurance, control, and management; organization's bidding procedures; handling of instructions and request for information; communication policies; project and contract reviews; project scheduling, administration, and management; project financial management including receipts and payouts as well as project controls; safety procedures; handling external constituencies like inspectors; handling incidents and emergencies; and deliveries and storages.

13.2 Getting Project Team Members to Anticipate and Proactively Manage Disputes

Every member of any project team should commit to PDM and get involved in a comprehensive review of any project from its bid to its final turnover. When invited to bid, or electing by itself to bid on a project, the firm should match such a project against its potential dispute metrics and flashpoints from start to finish in the following perspectives:

1) *Owner and representatives*: What do you know about this owner and his/her representatives on this project in question? What is their reputation with regard to disputes, and are they controversial, and have they been involved in concerning project disputes? These questions can be answered by deploying due diligence to investigate this owner and representatives, as well as the following:
   i) *Project scope*: Is the project scope one that is so vast and complex as to present significant challenges in managing associated dispute risks? This question can be answered by a comprehensive and thorough examination of the associated contract documents that include the program, drawings, specifications, addenda, bid instructions, contract, other preconstruction documents, amendments, and instructions. Are there observable ambiguities, discrepancies, omissions, and commissions? Are the roles and responsibilities of the parties and participants clearly defined and delineated? Can all these questions be satisfactorily answered before bid closing and selection of a winner? All hands must be on deck and pairs of eyes focused to proactively evaluate these perspectives in relation to potential disputes by all internal project team members.
   ii) *Observable triggers of disputes*: Are there observable triggers of disputes, and can they be addressed with one or more of the PDM precepts of D-PAD? Once more, all internal project teams are expected to participate in contributing to observing, isolating, evaluating, and determining any potential dispute flashpoints.
   iii) *Possibility to work with owner and representatives*: In the internal project team's assessment, does there appear to be a possibility of rapport and cooperativeness with the owner and representatives on the project? If the answer is in the affirmative from the internal members, then there's leeway to proceed with bid submission for the project with some certainty of successfully and proactively managing any associated dispute risks along the way with PDM precepts.
2) *Participating professionals and consultants*: Designate and deploy internal project team members to investigate and divulge disputatious red flags among

participating professionals and consultants to the project team. Besides design professionals and consultants working for the project owner such as architects and civil engineers, others include building services engineers (mechanical, electrical, communication, etc.), structural engineers, surveyors, quantity surveyors and/or cost engineers, program managers, construction or project managers, procurement managers, and estimators. Due diligence requires that all must be investigated for potential contributions to dispute risk. Using the DRA approach discussed in Chapter 11, the lead PDM or master captain should analyze the data gathered from team member investigations and match observable potential disputes to applicable PDM precepts of D-PAD, with the following additional considerations:

i) *Understand the roles and responsibilities of participating professionals and consultants*: Are their roles and responsibilities clearly delineated to avoid duplications and ambiguities? In order to answer this question, a thorough review of the agreements and contracts with which they are engaged is crucial. The designated internal team member(s) investigating each professional and/or consultant conduct the reviews and feed forward to the lead PDM who collects all the data gathered from every internal team member.
ii) *Evaluate communication and coordination protocols*: Ascertain whether there are set communication and coordination protocols among these professionals and consultants; identify if there are potential problem areas and aspects that need to be tidied up ahead of project commencement to prevent, avoid, or douse possible conflicts and disputes.

3) *Contractor and representatives*: This aspect involves internal team assessment of itself and the organization. What is our dispute reputation rating, and how have we managed conflicts and disputes in our projects over the years? Those contractor organizations that diligently document their conflict and dispute experiences as well as historical records can easily answer these questions and match them against the current project in view, with the hope of bettering their dispute management position over the past years.
4) *Subcontractors, suppliers, and vendors*: One of the tenets of the quality management movement is that organizations establish long-term relationships with others with whom they work, such as contractors with subcontractors, suppliers, and vendors. Hence, the subcontractors, suppliers, and vendors to be involved on a particular project are likely well-known by the contractor, and their dispute risks can be anticipated and isolated for appropriate treatment with the PDM precepts of D-PAD. The agreement and contractual clauses are known and might have been tested on a previous collaborative project. Communication and coordination protocols are already familiar, and existing relationships and knowledge of one another guarantee extra assurance and

confidence with regard to managing triggers of potential conflicts and disputes. While certain internal project team members are designated and deployed to certain subcontractors, suppliers, and vendors for thorough reviews or re-reviews, it is the duty of the plead PDM to gather and articulate all the data toward a seamless application of and matching to D-PAD.

5) *Contract documents*: Are the contract documents complete and not deficient in any way? Any embodied ambiguities or discrepancies in them? A comprehensive and thorough review and quality control and assurance of the contract documents ahead of construction start uncovers and likely adequately answers these questions. The documents should be reviewed vis-à-vis the project scope, key contract clauses, allocation of responsibilities, compliance, bonuses and penalties, precedents, communication protocols, and documentation. The master captain or the lead PDM should lead and articulate this aspect, even as all team members are expected to participate and contribute to combing through the contract documents for latent and potential triggers of conflicts and disputes against which D-PAD can be proactively applied.
6) *Location*: Leveraging on the outcomes of the project location's dispute risk analysis (DRA) discussed in Chapter 11, the internal project team can collate comprehensive, observable, potential conflict and dispute triggering factors. The lead PDM or master captain articulates the disparate factors and appropriately matches them to the PDM precepts of D-PAD. Such factors include site conditions, environmental and regulatory requirements, differences in people and cultures, labor and material supply chain, as well as accessibility, logistics, and roominess for staging. Matching these factors against D-PAD from the onset is a crucial step to proactively manage associated potential conflicts and disputes. DRA drives the ability of the internal project team and its lead PDM or master captain to appropriately and effectively accomplish this matching to proactively manage inherent potential conflict and dispute triggers.
7) *External factors*: External factors are project realities outside and beyond the control of the project team. As such, external constituencies may be involved, marshaling out all possible external factors portending potential conflict and dispute risks to the project in order to appropriately and proactively deploy D-PAD to checkmate them. These factors include legal, regulatory, economic, political, environmental, climate, community, and social influences on the project.

The project owner and representatives should be evaluated relative to compatibility and propensity toward dispute, reputation, and capital or financial capability—a diligent and comprehensive deployment of dispute risk analysis as discussed in Chapter 11.

In summary, the site should be investigated for potential issues, such as accessibility, logistics, local ordinances, labor availability, and nuisance from authorities and neighbors. If possible, the contract to use for the project should be examined and any potential disputatious aspects identified and clarified ahead of bidding. Any findings should be shared completely with the likely project team members.

The bidding stage presents an early opportunity to proactively eliminate potential disputatious issues by fielding and getting clarifications on ambiguities, discrepancies, and questions, and deploying dispute risk analysis (DRA) discussed in Chapter 11.

13.3 Designating One or Two Outstanding Persons as Lead PDM

In proactive dispute management, while all project team members are invited and encouraged to participate in driving and entrenching its practice in the process, lead proactive dispute managers lead the process. As earlier stated, while a ship may have more than one captain, there will always be one master captain responsible for the entire ship in the singular sense of ultimate command to avoid dangerous confusion. This scenario is applicable to PDM, where there may be two lead proactive dispute managers, but one must be the master captain with whom ultimate decisions reside under critical or crucial situations.

Designation of one or two lead PDMs or a master captain proceeds in one of many ways, namely:

1) *Designation by contracting construction firm as a whole*: Among its personnel, the firm may designate a staff who stands out as having an aptitude to effectively manage conflicts and disputes. This staff's designation as lead PDM or master captain may be a result of the simple majority of the firm's personnel supporting or voting for such staff for this role.
2) *Designation by application*: The personnel of the firm who deem themselves well equipped and fit for the lead PDM or master captain may apply and make cases for why they are most suitable for the role, and the rest of the personnel of the firm or representatives select the most compelling applicant for the role.
3) *Project team designation*: For a particular project, the team may designate the lead PDM or master captain based on the colleague's demonstrated and acknowledged aptitude and competence or knack for effectively managing conflicts and disputes. This designation can proceed by voting to enthrone a

sense of ownership of leadership by the team members. Nonetheless, the firm's top executives with whom the responsibility of all projects resides retain a veto power over the team vote in order to prevent the selection of an unsuitable but popular staff who may not be the most competent or effective at managing conflicts and disputes.

4) *Top management executive designation*: The firm's top management may handpick the lead PDM or master captain based on the staff's record and experience in managing conflicts and disputes. The top management's involvement in this process is to facilitate the selection of staff that best meet their PDM goals or objectives on the project, and to prevent the selection of an unsuitable but popular staff who may not be the most competent or effective at managing conflicts and disputes.
5) *Designation by project role*: Some firms may by default designate the lead project manager as also the lead PDM or master captain. The rationale for this designation is to avoid duplication and clashing of responsibilities, and the understanding that the PM holds the topmost responsibility to ensure the overall success of the project. Hence, the PM calls the final shots and is best positioned to oversee and proactively manage potential conflicts and disputes on the project.

Regardless of how the lead PDM or master captain is appointed, they must come with the qualities of (1) experience and reliability, (2) responsibility and effective communication ability, (3) servant leadership, and (4) role model persona. Their duties include:

1) *Proactively leading and motivating the project team members with regard to issues and triggers of conflicts and disputes.*
2) *Monitoring progress to ensure that the project goals are met and undisrupted by conflicts and disputes.*
3) *Serving as a point of contact or reference for all project stakeholders with regard to issues or potential triggers of conflicts and disputes.*
4) *Proactively communicating expectations and managing potential project conflicts and disputes by effectively applying the PDM precepts of D-PAD.*

13.4 Conclusion

PDM is new and calls for new approaches to avoid the pitfalls of existing means of handling project conflicts and disputes. While all project team members are invited and encouraged to participate in driving and entrenching its practice in the project delivery process, there must be a lead PDM or master captain responsible for the entire project ship in the singular sense of ultimate command to avoid dangerous confusions. The scenario prevents the situation of "too many cooks spoil the broth" with respect to project authority and responsibility for conflicts and disputes management.

With PDM discussed here, there's no outside consultant coming in to administer onerous workshops as in partnering. Rather, the PDM process is seamlessly woven into each project just at the time the project team is bidding or mobilizing to commence with the project. As a result, implementation of PDM on any project is relatively inexpensive, which encourages its integration in every project designated for success with literally no disruptive conflicts and disputes. Project team members continue to build experience by integrating and implementing PDM on one project after another until its adoption and practice become commonplace and universal.

Chapter Review Questions

1) A project can have multiple lead proactive dispute managers, but only one should be the master captain. True/False?
2) Lead PDMs are typically external consultants hired for each project. True/False?
3) Every project team member should be aware of and contribute to PDM practices. True/False?
4) The partnering approach was abandoned due to its cost and one-size-fits-all limitations. True/False?
5) PDM relies on external workshops and facilitators to manage disputes. True/False?
6) Designating a lead PDM can be done through application or team consensus. True/False?
7) The master captain in PDM has no authority over dispute decisions. True/False?
8) Which of the following is a method for designating a lead proactive dispute manager?
   a) Random selection
   b) External arbitration
   c) Application and team voting
   d) Government appointment
9) What was one major reason partnering was abandoned in construction dispute management?
   a) Lack of stakeholder interest
   b) No measurable outcomes
   c) Legal restrictions
   d) Excessive cost and rigid structure

10) What role does the master captain play in PDM?
    a) Oversees financial audits
    b) Handles all legal documentation
    c) Makes final decisions in critical dispute situations
    d) Manages subcontractor hiring
11) Which of the following is NOT a responsibility of a lead PDM?
    a) Managing project finances
    b) Promoting dispute awareness
    c) Coordinating dispute prevention strategies
    d) Leading internal dispute reviews
12) Which of the following best describes PDM's approach to dispute management?
    a) Reactive and litigation-based
    b) Proactive and internally driven
    c) Outsourced and standardized
    d) Passive and observational
13) What should be shared with all project team members to support PDM implementation?
    a) External arbitration contracts
    b) Organizational policies and procedures
    c) Partnering workshop schedules
    d) Litigation case studies
14) Which of the following is a key benefit of engaging lead PDMs?
    a) Reduced need for project documentation
    b) Elimination of all disputes
    c) Streamlined proactive dispute handling
    d) Increased reliance on external consultants
15) The lead proactive dispute manager responsible for final decisions is referred to as the ______.
16) PDM avoids the one-size-fits-all approach used in ______.
17) A lead PDM may be designated by firm-wide consensus or by ______.
18) Every project team member should be made aware of existing ______ procedures and policies.
19) The partnering model failed partly due to its reliance on ______ facilitators.
20) The master captain must be prepared to act during ______ dispute situations.

Exercises

Exercise 13.1 Lead PDM Selection Simulation

Objective:
Practice evaluating and selecting a lead proactive dispute manager for a construction project.

Instructions:

- Form a group and assume the role of a construction firm preparing to designate a lead PDM for a new hospital project.
- Each member will present themselves as a candidate for the lead PDM role, highlighting:
  - Relevant experience in dispute management
  - Understanding of D-PAD principles
  - Leadership and communication skills
- The group will evaluate each candidate based on a rubric (e.g., experience, team support, strategic thinking).
- Select one master captain and optionally a secondary lead PDM.
- Justify the selection and outline how the chosen PDMs will lead to dispute prevention, avoidance, and dousing throughout the project lifecycle.

Outcome:

Understand the criteria and process for selecting effective dispute leaders and the importance of internal leadership over external facilitation.

Exercise 13.2 PDM Integration Planning

Objective:

Develop a plan to integrate PDM leadership into a project team from day one.

Instructions:

- Choose a real or hypothetical construction project (e.g., an office building).
- Draft a plan for how the lead PDMs will be introduced and engaged during:
  - Preconstruction planning
  - Bid review and risk analysis
  - Team onboarding and training
  - Communication and reporting protocols
- Include strategies for ensuring all team members understand and support the PDM role.
- Present your plan as a short proposal or visual roadmap.

Outcome:

Learn how to embed proactive dispute leadership into project culture and operations from the outset.

Discussion

Prompt:

What other ways can someone show leadership in proactively managing project conflicts and disputes? Provide other examples of programs or initiatives adopting similar lead or master captain designations.

Prompt:

Provide examples of programs or initiatives adopting different lead or master captain designations, and discuss those different approaches. What may be the drawbacks of the different lead PDM or master captain designations offered here?

14

Implementing PDM Within an Establishment and Project

Readers' Learning Objectives and Outcomes

1) *Gain a clear understanding of how proactive dispute management (PDM) can be implemented at both the organizational level and the project level to reduce disputes and improve collaboration.*
2) *Learn practical methods for integrating PDM concepts–dispute prevention, avoidance, and dousing (D-PAD) into everyday workflows, contractual arrangements, and team practices.*
3) *Acquire the ability to identify different types of disputes and determine which PDM strategy is most suitable for addressing each situation to ensure smooth project delivery.*

14.1 Introduction

PDM is implemented at two levels, namely, (1) the establishment or firm level dealing mostly with dispute prevention structures, and (2) the individual project level concerning dispute avoidance and dispute dousing strategies. The reason is that conflicts and disputes are not exclusively domiciled in the projects, but they often have their foundations and origins in the establishment's structures, systems, practices, policies, and procedures. Hence, it is inadequate to wish to handle or manage them effectively at the project level without first putting the establishment's home in order with regard to preparedness to proactively manage conflicts and disputes. Doing so would be tantamount to a firefighting approach. Therefore, with regard to effective proactive management of project conflicts and disputes, charity must begin at home in this regard.

14.2 Implementing PDM at the Establishment Level

Implementing PDM at the establishment, firm, or organization level is all about dispute prevention. It involves businesses reviewing, revising, and buttressing their policies and procedures with the intent to prevent disputes, based on a reflective analysis of, and understanding from previous projects and the causes and triggers of disputes witnessed. As previously discussed in Chapter 1, dispute prevention is anchored and starts at the establishment or organization level, and it is associated with an organization's cultural initiatives, which guide the steps that are consciously and continuously taken by an entity to prevent conflicts and disputes from arising. The word "prevention" signifies an acknowledgement of the possibility of encountering latent and active conflicts and disputes in the process; hence, measures are taken to prevent as much of the potential of their occurrence as possible. Dispute prevention strategies include:

1) *Standing and strategic culture of working with comprehensive planning and detailed design*: The organization should set standards for projects that involve comprehensive planning and detailed design before commencement, rather than the adoption of design and detail as you go. Even in a fast-track delivery approach, emphasis must be on working with detailed design before commencing on any part of the project. Any ambiguities and discrepancies must be resolved ahead of project commencement. Such early efforts can proactively set the pace to prevent conflicts and disputes as the project delivery is in progress.
2) *Adequate and effective communication*: Adequate and effective communication is known to promote stronger team collaborations and relationships, enhanced participants and stakeholders' engagement and satisfaction, increased productivity from better and more efficient task prioritization and resource allocation, timeliness, and better cost and quality management, and ultimately minimization or elimination of conflicts and disputes.
3) *Working with unambiguous and explicit contracts*: Insisting on working with clearly defined, unambiguous, and easily or well-understood contract clauses eliminates guesswork and possible misinterpretations that can trigger conflicts and disputes as the project progresses. Doing so can minimize or eliminate costly conflict and dispute risks, improve relationships by building trust and setting clear expectations and communications, increase efficiency by saving research and interpretation time, and offer greater enforceability if and when the need arises.
4) *Practicing proactive dispute risk analysis (DRA) and assessments of projects*: This practice allows organizations and their project teams to timely identify, isolate, evaluate, and mitigate potential conflict and dispute triggers ahead of any project commencement. This forward-thinking approach offers many benefits that include cost and losses control and inefficiencies associated with potential conflicts and disputes, better participants and stakeholders' collaboration and relationships, and ultimately a better guarantee of project success.
5) *Imbibing of total quality management, quality assurance, and quality control (TQM/QA/QC) principles*: These principles benefit projects by improving service and outcome quality, optimizing cost, and minimizing or eliminating waste, boosting efficiency and productivity, while fostering a culture of maximizing customer satisfaction and continuous overall process improvement. TQM offers a comprehensive, tactical, and strategic approach to overall quality matters involving all the organization's personnel, while QA

establishes the processes and metrics for achieving overall quality, and QC ensures that the final project outcomes and deliverables meet the set quality metrics and standards.

6) *Adoption of Six Sigma principles*: Six Sigma is simply the raising of quality assurance to the statistical sixth sigma level; that is, driving toward six standard deviations between the mean and the nearest specification limit. Originally developed by Motorola in 1986, Six Sigma stands to offer project or product process outcomes in which 99.99966% are statistically expected to be free of defects in the order of no more than 3.4 defects per million parts. It is therefore a higher-order quality assurance. The two Six Sigma sub-methodologies are (1) define, measure, analyze, improve, control (DMAIC) for existing product or project processes needing incremental improvements, and (2) define, measure, analyze, design, verify (DMADV) for new product or project processes seeking to proceed at Six Sigma quality levels. Both sub-methodologies' implementations are led and executed by Six Sigma Green Belts and Six Sigma Black Belts, and they are overseen by Six Sigma Master Black Belts. According to the Six Sigma Academy, Black Belts have been touted to have saved companies approximately $230,000 per project. General Electric, which adopted Six Sigma in 1995 after Motorola and Allied Signal blazed the trail, has reported benefits in the order of $10 billion during the first five years of implementation. Since then, many companies and businesses around the world have adopted Six Sigma principles and testify to the immense benefits.
7) *Adherence to dispute reputation rating (DRR) metrics*: Just like experience modification rating (EMR), aspiring to achieve a better DRR provides significant benefits in minimizing or eliminating disruptive conflicts and disputes on projects, minimizing or eliminating conflict and dispute costs on projects, making the firm more attractive to owners and more competitive at winning projects, and demonstrating a commitment to peace and tranquility through project delivery that attracts and retains more project owners' patronage that promotes business continuity and survival. A firm with a poor DRR will not survive. Hence, having robust means and metrics to monitor and minimize or eliminate conflicts and disputes that call for resource waste on projects are sure way for a firm's business vitality, survival, and success.
8) *Openness to collaborative approach and effective document handling, distribution, and interpretation*: Openness to collaborative approach (such as alliancing, integrated project delivery, and partnering) and effective document handling, distribution, and interpretation culturally align with effective PDM implementation at the organization level.
9) *A culture of comprehensive and thorough design review and value engineering*: The benefits of a culture of comprehensive and thorough early design

review and value engineering to PDM implementation at the organization level cannot be overstated, especially with regard to project constructability analysis to hash out and clear all construction bottlenecks. These are potent measures in the dispute prevention toolbox. Never again build new or renovate projects without first conducting a comprehensive design review and value engineering. This process guarantees the best value for every dollar invested in the project without compromising quality, and it also improves functionality and performance, fosters innovation, promotes sustainability and efficient use of resources, waste minimization or elimination, and energy conserving and efficient practices. All of these benefits lay the foundation for kick-starting owner, stakeholders, and participants' satisfaction with the project process, which would translate to the right atmosphere to mitigate and/or eliminate conflicts and disputes in the project delivery process.

10) *PDM Memorandum of Understanding (MOU)*: Sealing the project contract (no matter the type) with an oversailing PDM MOU that supersedes and takes precedence over differing contract clauses. Just like some contracts stipulate that disputes be submitted to mediation or arbitration rather than litigation, a PDM MOU can specify that parties to the contract submit to, and implement, PDM's DPAD in exclusion of ADR and/or litigation. Both parties should commit to this PDM MOU from the start. While the PDM MOU instrument is operational at the project level just like a special or supplementary condition of contract, its adoption and institutionalization are at the establishment, speaking to its operational culture or policies on projects. Hence, the PDM MOU can rightly represent PDM implementation at both the establishment and the project levels.

14.3 Implementing PDM at the Project Level

Implementing PDM at the project level involves the second and third concepts or precepts of PDM, namely, dispute avoidance and dispute dousing.

1) *Dispute avoidance*: As previously presented in Chapter 1, dispute avoidance proceeds with the elephant in the room analogy and refers to actions to keep or navigate away from extant potential conflict and dispute flashpoints and not do anything to agitate them in the project delivery process. Due to the ever presence of Murphy's Law, even with our best intentions and efforts, things that can go wrong will go wrong. The word "avoidance" suggests an acknowledgement of the extant latent and possibly active conflict and dispute flashpoints in the project delivery process; hence, steps are taken to timely navigate around them and eliminate disruptions from them. Dispute avoidance strategies include both a priori and a posteriori knowledge and experience in

the administration and management of the construction process. Projecting and envisioning potentially disruptive aspects of the project, based on accumulated previous experience, gives the ability to timely navigate around them and keep the work moving forward with minimal or no disruptions that can avoid conflicts and disputes. Concrete work and its setting and curing time were already given as an example of an aspect of some projects susceptible to disrupting the schedules of other trades and that of the project as a whole, potentially being a flashpoint for conflicts and disputes. Based on experience, several aspects of a typical project portend flashpoints for potential conflicts and disputes, depending on type and scope. Excepting aspects where this distinction appears blurred and indistinct between dispute prevention and dispute avoidance, Morgan (2008) offered several plausible dispute avoidance techniques at the project level that include:

a) *Know your project owner, representatives, and other participants through the following:*
   i) *Due diligence investigation*: Due diligence investigation and understanding of the project owner, representatives, and other participants when invited to bid are the first timely steps to avoiding disputes as the project progresses. A comprehensive investigation based primarily on the commercial and legal appraisal status of the owner is in order. Discussions with local subcontractors, suppliers, and vendors who have previously worked with the owner can X-ray its approach to payment, ethics, treatment of others, and relationships with relevant authorities, etc., which will all have a significant bearing on the successful delivery of the project being bid with minimal or no hitches or conflicts and disputes.
   ii) *Leverage on previous experience of the project owner*: It may be possible, especially for medium or large firms, to have an internal staff who might have had previous involvement with or experience of the project owner, consultants, partners, and participating subcontractors, suppliers, and vendors. That personnel can be enlisted to unravel the owner's business outlook and ethics. The bidding firm may, through this personnel, access dispute records of the owner when he/she was last involved with this party, why these disputes arose, and how they were dealt with. Any previous knowledge of the project owner held by the bidding firm that might assist the understanding of this other party should be passed on at the earliest opportunity.
   iii) *Maintain cooperative relationships with the project owner and other participants*: Maintaining such relationships facilitates familiarity or understanding of the other party's and participants' strengths and weaknesses, which will make it possible to work with them in a way that can build mutual trust and cooperation. Emphasizing the positive aspects or attributes of the other party and participants will facilitate

closer cooperation and better interaction with them, thus mitigating and enabling the avoidance of conflicts and disputes among them.

b) *Know, train, and support the staff involved in the project by doing the following*:
   i) *Understand project staff strengths and weaknesses*: Selection of project team members should not be merely based on availability, but on the capability and previous residual experience to effectively handle the current project at hand. To be taken into consideration is the manner in which each project team member is likely to interact with each other, as well as the other parties' and participants' teams, and the local population around the project site.
   ii) *Staff positioning*: There may be a staff that must be included in a particular project for political or internal firm's reasons as the right front for the firm on the project and its location. This staff may be the one possessing the skills of presentation, negotiation, and persuasion; it is understandable that the best staff to research and collate data may not be the best to present that data to others.
   iii) *Human or interpersonal skills*: For each project, involve staff with human or interpersonal skills. Although human skills and charisma are congenital and cannot be taught, improvements can be made through dedicated mentoring for those who don't possess them. Nonetheless, there is no doubt that acquired skills are never truly as good as those that come naturally.
   iv) *Directed training of the firm's staff in dispute management*: Morgan (2008) also recommended having dispute management staff and directionally training them to have a full and complete understanding of, and acceptable level of knowledge in, the core competencies required of someone in their particular role on the project at hand. The core competencies required of a dispute management team member are:
      I) *Knowledge*: Knowledge here includes ability to understand behavioral aspects of dispute; cross-cultural consideration; casualties of disputes such as loss of productivity and low morale; different approaches to dispute management and their efficacies; basic motivation theories; power dynamics and the impact on the parties of imbalance of power; dispute cycles from emergence to escalation and deescalation; differences between roles, process, responsibilities, authorities, duties and or tasks; differences between compromise, cooperation, collaboration, and consensus building; and personality and dispute handling styles, strengths and challenges.
      II) *Skills*: Every team member must demonstrate collaborative problem-solving skills in active listening; formulating and expressing strategies defining the required outcomes; identifying underlying

interests; ascertaining, developing, and analyzing options; and identifying different courses of action and analyzing and understanding the consequences and relative benefits of each, including actual and consequential costs and risks.

III) *Attitude*: Every team member must demonstrate that they have an understanding of ethical behavior required of them in a collaborative dispute management context that includes neutrality; confidentiality; objectivity; respect for differences among parties and participants; honesty and credibility; professional accountability and integrity; and impartiality.

v) *Danger signals*: Those aspects of the project that are prone to conflicts and disputes tend to send danger signals as they incubate, and they need a team with experience to artfully and timely navigate around them by avoiding to agitate or detonate them.

Morgan (2008) offered many more techniques, which for brevity are not elaborated further upon here, and these include: understanding the contract being used and any latent or inherent potential for conflicts and disputes; committing to abiding by the letter of the contract clauses and terms with avoidance of conflicts and disputes a top priority; conducting comprehensive and thorough DRA way ahead of project commencement; deploying effective and timely open communication; active listening; and having procedural guidance in place.

2) *Dispute dousing*: As previously presented in Chapter 1, dispute dousing is engaged when dispute prevention and dispute avoidance fail, and it involves the deployment of quick actions to neutralize and extinguish emergent conflicts and disputes before they establish, deteriorate, and escalate into costly and time-consuming ADR or litigation. The word "dousing" suggests that the fire of conflicts and/or disputes is about to ignite, and quick steps must be timely taken to quench or extinguish it before it establishes and grows to a point that you have to fight it with something more than a simple fire extinguisher. Dispute dousing can be achieved by continuous close monitoring of the process to ensure early detection of budding dispute issues, open communication channels with the parties and participants, encouragement of collaborative solutions to emerging issues, and clear task and responsibility assignments and leadership. Charisma and probably some conflict handling and resolution training, as well as intermittent and regular audits and reviews, are helpful within the dispute dousing toolbox. In order to effectively deploy dispute dousing, the following steps are imperative:

a) *Before construction starts*:

i) *Thorough DRA*: Please refer to Chapter 11 on DRA. The DRA must involve a deconstruction and reconstruction of the project model and its delivery from the construction documents by the project team, with an eye to identifying potential conflict and dispute triggers and how to handle them. A precise summary of the DRA analysis outcomes must be openly shared and communicated to other parties to the building contracts.

ii) *Working with clear and comprehensive contracts*: Ambiguous and poorly drafted contracts are potential triggers of discrepancies, disagreements, conflicts, and disputes during construction delivery. Hence, ensuring the definition of the scope of work, schedules, payment terms, responsibilities delineation, and assignments, as well as risk allocation, are key in a comprehensive and unambiguous contract.

iii) *Working with realistic schedules and expectations*: A mismatch between the schedules and the work scope can conflict with expectations and become a potential trigger of disagreements, conflicts, and disputes during construction delivery. Hence, avoiding aggressive but unrealistic timelines and prices without allowing for possible delays and uncertainties is key in taming potential conflict and dispute triggers that may need dousing during construction delivery.

iv) *Establishing clear communication lines and protocols early*: Defining a chain of command as well as channels and lines of communication from the onset sets the foundation for clear and direct communication, ensuring alignment and timely information for every project participant and stakeholder.

v) *Ensuring that project participants can competently perform their roles*: Carefully selecting competent project participants, such as subcontractors and vendors, more readily guarantees that performance issues are mitigated, as are conflicts and disputes that can be triggered by them.

vi) *Engaging technology in streamlining and managing project communications and documentations:* Deploy project management software and tools that can help centralize and streamline project communications and documentations.

b) *During construction delivery*:

i) *Meticulous documentation:* Keep a detailed record of all communications, deliveries, visitations, inspections, including daily job logs and reports, timestamped photos, emails, meetings, and minutes. Having these documentations available and accessible for timely reference as and when needed is a major strategy in dousing potential conflicts and disputes.

ii) *Clear, timely communication:* For any latent, anticipated, or potential issues, such as material or labor shortages, promptly notify all participants and stakeholders that may be affected, present navigable options, and document the collective or consensus decisions.

iii) *Formally manage changes:* Change or variation orders should be appropriately and formally documented as such, most especially because they modify the contract sum and schedule—a potential conflict and dispute trigger. There should be a clear distinction between change orders and field orders, with the latter having no impact on modifying the contract sum or schedule. Ensure that the parties sign off on both.

iv) *Deescalate emotions and tensions*: It can be expected that emotions and tensions may sometimes arise when there are many participants on a project with different daily stories and challenges. Figuring out a way to deescalate emotions and tensions among project participants is a key asset on any job. One such crucial strategy is keeping calm and professional in dealing with issues and individuals, focusing on facts and the best interests of the project and the participants. Even calling a short timeout can sometimes do the miracle.

v) *Foster collaboration rather than individualism*: A collaborative rather than individualistic atmosphere promotes camaraderie as well as shared responsibility among participants for the success of the project. Mutually beneficial solutions are embraced under this milieu as opposed to blame games that can be rife with individualism.

vi) *Strict interpretation and compliance with the contract*: One may never see any latent or potential conflicts and disputes to douse if the contract is strictly interpreted and followed in the execution of the project, especially where the contract embodies no ambiguities and discrepancies. Adhere to the contract clauses and terms on interim or periodic payments, change orders, delays, time extension, and damages—liquidated or ascertained.

14.4 Matching Dispute Types with PDM Concept Types: D-PAD

Following the aforementioned and discussed dispute triggers and their broad categorizations, their individual PDMs' D-PAD concept matches are shown below in Table 14.1.

14.5 Conclusion

In summary and reiteration, most experienced industry professionals are aware of what leads to disputes or potential triggers and areas that are susceptible to disputes, or what can generate disputes. It means that most of these disputes are predictable, and measures can be put in place to prevent or avoid them, and if both steps fail, then swift action is taken to douse them quickly before they escalate

Table 14.1 Matrix matching dispute triggers to PDM concept types.

Dispute, D-	P	A	D
Owner-related dispute flashpoints or triggers			
Addenda		✓	✓
Site conditions	✓	✓	✓
Change order or variation order by owner		✓	✓
Scope creep		✓	✓
Project turnover or handover hitches	✓	✓	✓
Constructive acceleration, lofty and unrealistic expectations by owner	✓	✓	✓
Payment delays	✓		✓
Contractor-related dispute flashpoints or triggers			
Financial capacity	✓		
Progress delays	✓	✓	✓
Liquidated and ascertained damages	✓		
Progress or interim payments and their misuse	✓	✓	✓
Character and personality of contractor		✓	✓
Technical capability and experience	✓		
Labor productivity fluctuations		✓	✓
Equipment productivity fluctuation		✓	✓
Inclusion of alternates	✓	✓	✓
Underbidding and accuracy of estimates	✓	✓	✓
Unrealistic scheduling	✓	✓	✓
Project resourcing and balancing—labor, materials, and equipment	✓	✓	✓
Use of union and nonunion labor		✓	✓
Work quality—omissions, defects, failures, rework, and punch-list	✓	✓	✓
Workers' fidelity	✓	✓	✓
Accidents and safety—experience modification rating	✓	✓	✓
Mechanics lien		✓	✓
Bonds	✓	✓	✓
Insurance	✓	✓	✓
Subcontractor and supplier-related dispute flashpoints or triggers			
Underbidding and equitable estoppel	✓		
Conflicting work schedules and coordination	✓	✓	✓

(Continued)

Table 14.1 (Continued)

Dispute, D-	P	A	D
Construction documents-related dispute flashpoints or triggers			
Quality of drawing documents	✓	✓	✓
Quality of specifications	✓	✓	✓
Discrepancies	✓	✓	✓
Reading and interpretation of construction documents	✓	✓	✓
Contract-related dispute flashpoints or triggers			
Ambiguities in the contract	✓	✓	✓
Discordant understanding and interpretation of contract clauses	✓	✓	✓
Omissions and commissions	✓	✓	✓
Sharing of risks and rewards	✓	✓	✓
Attitude and disposition-related dispute flashpoints or triggers			
Litigious nature of parties and participants	✓	✓	✓
Cultural disposition to work	✓	✓	✓
Cultural disposition to contracts	✓	✓	✓
Communication and miscommunication	✓	✓	✓
Many players on the one project	✓	✓	✓
Team-playing	✓	✓	✓
Nature and uniqueness of project dispute flashpoints or triggers			
Locational disparities and site conditions	✓	✓	✓
Novelty of project	✓	✓	✓
Rate of request for information	✓	✓	✓
Poor detailing of novel structures	✓	✓	✓
Unforeseen circumstances	✓	✓	✓
External factors-related dispute flashpoints or triggers			
Weather conditions		✓	✓
Force majeure		✓	✓
Regulatory enforcements		✓	✓
Politico-economic changes		✓	✓
Differential lead times	✓	✓	✓
Labor unions	✓	✓	✓
Labor seasonality	✓	✓	✓

P = Prevention, **A** = Avoidance, **D** = Dousing.

to a degree that requires resource-consuming resolution or litigation. Hence, identifying, isolating, and managing them early ahead of possible occurrence are the efficacious steps to PDM and its concepts or precepts of D-PAD. This purpose is achieved through early internal DRA before commencement of the construction project. Much like TQM, quality assurance and control on any job, engaging in an internal PDM process prior to construction can avoid third-party involvement such as partnering or alliancing, ADR, and litigation, thus eliminating expense, lost time, owner involvement, and risk of loss of reputation and future business.

Reference and Suggested Further Reading

Morgan, D.B. (2008). *Dispute Avoidance: A Non-confrontational Approach to the Management of Construction Contracts*. RIBA Publishing, London.

Chapter Review Questions

1) PDM can only be implemented at the project level, not at the establishment level. True/False?
2) Matching dispute types with D-PAD concepts is a key part of implementing PDM. True/False?
3) Once implemented, PDM does not require continuous monitoring or improvement. True/False?
4) Implementing PDM helps reduce reliance on third-party dispute resolution methods like ADR and litigation. True/False?
5) Dispute dousing should only be applied after disputes escalate. True/False?
6) A robust monitoring and reporting system is essential for effective PDM implementation. True/False?
7) PDM implementation is most effective when integrated early in the project lifecycle. True/False?
8) Implementing PDM at the organization level involves ______________.
   a) Dispute prevention
   b) Dispute avoidance
   c) Dispute dousing
   d) All of the above
   e) None of the above
9) Implementing PDM at the project level involves ______________.
   a) Dispute prevention
   b) Dispute avoidance
   c) Dispute dousing

   d) a and b
   e) a and c
   f) b and c
   g) All of the above
   h) None of the above
10) Which of the following is a core strategy in implementing PDM at the establishment level?
   a) Outsourcing dispute resolution
   b) Creating a culture of dispute prevention
   c) Avoiding documentation
   d) Relying solely on litigation
11) Which of the following is NOT a recommended practice for implementing PDM at the project level?
   a) Matching dispute types with D-PAD
   b) Monitoring dispute risks
   c) Engaging lead PDMs
   d) Ignoring feedback after project completion
12) What is the primary goal of implementing PDM within a project?
   a) To reduce the number of subcontractors
   b) To eliminate all contracts
   c) To proactively manage and reduce disputes
   d) To increase litigation opportunities
13) Which of the following best describes dispute avoidance in the D-PAD framework?
   a) Ignoring potential issues
   b) Navigating around known dispute flashpoints
   c) Waiting for disputes to escalate
   d) Hiring a legal team
14) Which of the following is a benefit of implementing PDM across an organization?
   a) Increased reliance on arbitration
   b) Reduced project transparency
   c) Improved dispute handling culture
   d) Elimination of all project risks
15) What should be done when dispute risks cannot be prevented or avoided?
   a) Apply dispute dousing strategies
   b) Escalate to litigation immediately
   c) Ignore the issue
   d) Delay project progress

16) A proactive dispute management system should include regular ______ and reporting.
17) Implementing PDM at the project level involves matching dispute types with ______ concepts.
18) Postconstruction ______ and review are essential for improving future PDM strategies.
19) A robust ______ system helps track project progress and identify emerging dispute risks.
20) PDM implementation encourages early identification and ______ of potential dispute risks.

Exercises

Exercise 14.1 PDM Organizational Culture Audit

Objective:
Evaluate how well an organization's culture supports proactive dispute management.
Instructions:
- Choose a real or hypothetical construction company.
- Conduct a mock audit of the organization's culture using the following criteria:
  - Leadership support for dispute prevention
  - Communication transparency across departments
  - Historical reliance on ADR or litigation
  - Employee training in D-PAD principles
  - Internal reporting and feedback systems
- Rate each category on a scale of 1–5 (1 = poor, 5 = excellent).
- Identify gaps and recommend specific cultural shifts or initiatives to support PDM implementation.

Outcome:
Understand how organizational culture influences the success of PDM and identify actionable improvements.

Exercise 14.2 Dispute Type–D-PAD Matching Workshop

Objective:
Practice aligning real-world dispute types with appropriate D-PAD strategies.
Instructions:
- Create a list of 10 real or hypothetical dispute scenarios (e.g., payment delays, scope creep, miscommunication, regulatory enforcement).

- **For each scenario:**
  - Identify the root cause of the dispute.
  - Determine whether it is best addressed through **prevention**, **avoidance**, or **dousing**.
  - Justify your choice with a brief explanation.
- Present your findings in a group discussion or visual chart.

Outcome:

Build fluency in applying the D-PAD framework to diverse dispute situations and reinforce strategic thinking in PDM deployment.

Discussion

1) What are the challenges of implementing PDM as a standard practice across an entire organization? Discuss how leadership, training, and internal systems can support or hinder the adoption of proactive dispute management.
2) How does implementing PDM at the project level differ from implementing it at the establishment level? Explore the benefits and limitations of each approach and how they can complement each other.

15

Dispute Reputation Rating (DRR) and Dispute Recordkeeping

Readers' Learning Objectives and Outcomes

1) *Understand the concept of dispute reputation rating (DRR) and why it matters for businesses, and how it influences competitiveness and client trust in the industry.*
2) *Learn practical methods for establishing DRR metrics, tracking dispute histories, and creating transparent recordkeeping processes that support proactive dispute management.*
3) *Identify how DRR can serve as an intervention mechanism to promote a proactive dispute management culture, improve reputation, and minimize costly dispute resolution processes.*

15.1 Introduction

While negotiated, mediated, arbitrated, and litigated conflicts and disputes abound in construction and related industries and waste significant amounts of resources each, they have been accepted as a bona fide part of the business process. Bidders and contractors are continually asked the following uncomfortable questions such as, "Are there any judgments, claims, arbitration proceedings or suits pending or outstanding against your organization or its officers?" and "Has your organization filed any lawsuits or claims with regard to construction contracts within the last five years?"

These are questions frequently asked by our owners during the RFQ/RFP stage as is typical in our industry today.

1) Please describe any lawsuits that were filed against your company in the last five years and the results of those lawsuits. Please describe any mediation or arbitrations your company has been involved with in the last five years and the results of those arbitrations/mediations.
2) Disclose all enforcement actions by professional licensing boards, courts, and other bodies, along with other matters, which may reflect upon the Firm's (or its proposed partners') professional and/or other qualifications. Describe any pending litigation against the Firm or its partners and any other factors that could affect the Firm's ability to perform and complete the Project.
3) Has your organization ever defaulted on a contract? Yes No.
4) Are there any judgments, claims, arbitration proceedings, or suits pending or outstanding against your organization or its officers? Yes No.
5) Has your organization filed any lawsuits or claims with regard to construction contracts within the last five years? Yes No.

6) Has your firm or any of its officers been involved in any lawsuits, arbitration, mediation or similar proceedings concerning construction contracts within the last five years? If the answer to this question is yes, please explain.
7) Has your organization filed any law suits or requested arbitration with regard to construction contracts within the last five years?
8) If the answer is yes to any of the above questions, please provide details and include in Attachment I -Supplemental Information.

Still, little or no effort has been made to activate a process similar to experience modification rating (EMR) to proactively and continuously monitor the incidents of resource-wasting negotiated, mediated, arbitrated, and litigated conflicts and disputes with the intent to minimize or eliminate them and improve their optics in this regard. Hence, it is timely and rational to introduce a process that systematically documents negotiated, mediated, arbitrated, and litigated conflicts and disputes to compute and rate a business or organization's dispute involvement and reputation under the dispute reputation rating (DRR) nomenclature. DRR continuous documentation, monitoring, and improvements, much like EMR, provide a ready reference for owners, thus removing the discomfort out of these questions if there is any need for them to still exist.

15.1.1 Weaning Businesses off ADR and Litigation

It is long overdue to wean businesses off resource-wasting ADR and litigation, namely, negotiated, mediated, arbitrated, and litigated conflicts and disputes, especially in construction and related industries. Reiterating the point made in Chapter 1, construction and related industries should be weaned off acceptance of and dependence on conventional processes of negotiation, mediation, arbitration, or litigation to which they have always run for ages to resolve preventable, avoidable, and dousable disputes. The key reasons for this need include:

1) *Expensive process with significant financial risk*: In monetary terms, negotiated, mediated, arbitrated, and litigated conflicts and disputes are expensive to the parties involved, and the order rises from negotiation to mediation, arbitration, and ultimately to litigation. For mediated, arbitrated, or litigated conflicts and disputes, expert witnesses are brought on board with substantial charges and fees that could have been dedicated solely to core, successful project delivery. Oftentimes, there's little allowance made for these expenses in the construction contracts and sums, and this can make a difference in whether a project is profitable or not.
2) *Loss of productivity*: The processes are time-consuming and call for redirection of time resources to issues of conflicts and disputes rather than to the actual,

timely project delivery. Time is spent gathering and providing documents for evidence at associated meetings.

3) *Strained relationship*: Even when conflicts and/or disputes are settled by these means with a winner and loser or a compromise, bad blood can linger between the parties through the process and after. The relationship between them can be strained to the extent of affecting future transactions or the possibility of one ever taking place between them.
4) *Building of bad reputation*: Even when conflicts and/or disputes are settled by these means, neither party comes out with their reputation intact with regard to the dispute. For those in construction and related industries with small circles where interpersonal relationships matter much, a business can easily find itself one or a couple more mediation/arbitration/litigation incidents away from closing shop for good.
5) *Uncertainty of winning recovery and hollow judgment or verdict*: The result of the process could be a win or a loss for either party. To the party losing, recovery is zero, and the cost of the process is a loss. To the party winning, the costs may not justify the recovery, what with other unsavory consequences impinging on notoriety and unwelcome reputation, as well as strained relationships. Even where the outcome is that both parties compromised, the remedy may not be worth the effort. For litigations, the judgment could be hollow for the winning party if the losing party declares bankruptcy. Whichever way, the outcome is never optimal for both parties, which is the critical imperative of PDM with proactive dispute prevention, avoidance, and dousing.

There's a recognition that the tendency for conflicts and disputes varies significantly among industry players, and they should not be put in the same box. The players must be distinguished, and those doing better in conflict and dispute frequency and management should be easily identified and rewarded, while denying such to others with unenviable records, with the aim to incentivize everyone to raise their efforts to better their image in this regard. Systematically documenting negotiated, mediated, arbitrated, and litigated conflicts and disputes is therefore imperative and long overdue to justify DRR for construction and related industries.

15.1.2 DRR as Intervention Vehicle for PDM

Besides efforts under PDM to proactively prevent, avoid, and douse conflicts and disputes, continuously and meticulously documenting their occurrence and costs provides a visual appreciation of their enormity, thus potentially motivating a dedication and determination to minimize and/or eliminate them to save the parties' costs, maintain amicable working relationships, and improve and protect their reputation in terms of conflict and dispute optics and image. This is the juncture where DRR comes in as a robust intervention vehicle to bolster PDM. When DRR is taken over and centrally administered by the owner or contractor organization or an independent institution such as the National Council for Compensation Insurance (NCCI) with regard to EMR, its use and application will become systematic for the parties. Owners will cease to ask bidders and contractors those apprehensive, multifarious questions relating to previous involvement in resource-wasting negotiation, mediation, arbitration, and litigation. Sometimes, these questions are open-ended with time limitations.

DRR is a worthy vehicle for PDM due to the following key reasons:

1) *Cost saving*: Documenting the occurrence of negotiated, mediated, arbitrated, and litigated conflicts and disputes does not constitute unexpected additional costs as the cost of negotiation, mediation, arbitration, and litigation processes.
2) *Ease of documentation*: Negotiated, mediated, arbitrated, and litigated conflicts and disputes can be easily and effortlessly documented as the project progresses through delivery. Any project team member can be designated to document and file the occurrence.
3) *Self-evaluation and reflection:* Documenting the occurrence of negotiated, mediated, arbitrated, and litigated conflicts and disputes can constitute a crucial self-evaluation and reflection exercise that can endear contractors, subcontractors, etc., to prospective clients and owners. Doing so speaks to the due diligence of the business in matters as important as handling conflicts and disputes in the business process.

15.1.3 DRR Metrics

Just as EMR, DRR metrics will evolve over time, but the seminal articulation is as follows:

DRR will be based on the three years prior to the most recent expired event. That is, the rate stays with a company for three years. New companies will start with a zero [0] DRR. Rating is based on the number and severity of the disputes in terms of cost and time. For moderation to make it equitable across small, medium, and large firms, the rating is relativized to the yearly number of projects (residential, commercial, industrial, heavy/highway/civil engineering/infrastructure), size/scope (residential (SQFT/SQM), commercial (SQFT/SQM), industrial (SQFT/SQM), engineering/infrastructure (SQFT/SQM/SQYD)), and total budget or construction costs of the projects, or simply annual revenue. The number of employees is discounted, but small companies are those that have 1–50 employees, medium-sized companies have 51–250 employees, and large companies employ over 250 people.

The U.S. Small Business Administration (SBA) sets official size standards for construction companies based on North American Industry Classification System (NAICS) codes for federal contracting. As adjusted in 2023:

1) Small companies are those with an average annual revenue of $45 million or less (it was $40 million before 2023).
2) Medium-sized companies, though not officially categorized, fall within an average annual revenue of $45–$250 million.
3) Large companies are those receiving an average annual revenue of over $250 million.

Subject evolving revisions, below are numbers assigned as raw, unmoderated scores for incidents of negotiation, mediation, arbitration, and litigation, respectively:

0.5 for conflicts and disputes that didn't rise up to mediation, and these include recorded or gazetted private negotiations requiring investment of resources of time and money.
2 for each mediation.
2.5 for each arbitration or mini-trial.
5 for each litigation.

Based on the size of the company (small, medium, or large), moderation is given as the total number of incidents for the three years prior divided by 1 for small companies, the total number of incidents for the three years prior divided by 2 for medium-sized companies, and the total number of incidents for the three years prior divided by 3 for large companies.

For the three years prior to the most recent expired event:

0–1 DRR—very low
1.5-4—low
4.5-7—medium
7.5-10—high
Above 10—extreme.
Less than 1 DRR is the most desirable.

Implications for companies: Bonds like payment and performance bonds, and insurance like builder's risk.

15.1.4 DRR Benefits

DRR benefits the parties, participants, and stakeholders in construction and related industries by documenting, tracking, and X-raying the construction companies' disputatious dispositions at a glance. A good DRR would enhance a company's reputation and competitiveness among potential project owners, participants, and stakeholders. Here are the specific benefits to each constituency:

1) For contractors/employers:
   a) *Competitive edge*: A good DRR translates to the optics of less disputatious disposition that can endear the company to most prospective project owners over their competitors. Owners may begin to set thresholds for bidding to a certain DRR for potential bidders, and those with a DRR higher than the threshold are denied bidding opportunity.
   b) *Possible less distraction and timely project delivery*: With a good DRR, there is increased assurance that project delivery will not be disrupted or hampered by conflicts and disputes that inevitably consume resources.
   c) *Continuous self-monitoring and improvement*: Continuously documenting and tracking incidents or conflicts and disputes that consume resources is due diligence on the part of the company, offering it multiple benefits to its bottom line.
   d) *Possible lowering of bonding and insurance premiums*: A good DRR can potentially lower bonding and insurance costs, and contested conflicts and disputes are part of their investigation and questions to companies coming to them for bonding and insurance.
   e) *Dispute risk management*: DRR is a bona fide instrument of proactive dispute management, as the scores bolster efforts at proactively identifying, isolating, and actively or effectively preventing, avoiding, and dousing project conflicts and disputes.
2) For project owners:
   a) *Reduced or minimal project risk from conflicts and disputes:* A good DRR assures project owners of the contractors' minimal to nonexistent disputatious tendencies. It assures them of project delivery that may be least distracted by resource-wasting conflicts and disputes.
   b) *Assurance on cooperativeness of the contracting company*. With a good DRR, project owners can be assured of the contractor's cooperativeness and likely more amicable dealings and relationships throughout their project delivery.
   c) *Better performance and more successful projects:* With better cooperativeness and minimal or no presence of conflict and dispute distractions, the contractors are likely to perform better and more successfully and timely deliver projects to both parties' financial and relational benefits.
3) For other project participants and stakeholders:
   a) Other project participants and stakeholders, such as employees, subcontractors and suppliers, regulatory bodies, banking, bonding, and insurance institutions, all benefit from the minimization or absence of conflicts and disputes in the project delivery process. Work progresses smoothly for all with minimal or no distractions from conflicts and disputes, and some

project risks are reduced or eliminated in this aspect. With DRR, banking, bonding, and insurance providers can easily and more accurately or systematically assess risk aspects to do with contested conflicts and disputes. Loss, claim, and settlement frequencies from contested conflicts and disputes would drastically reduce when their customers are more conscious of this aspect through continuous documentation and monitoring for DRR.

15.2 Conclusion

DRR is crucial and a boon to achieving minimal or no conflict and dispute distractions in project delivery. It provides a hitherto inexistent metric to monitor and minimize or reduce contested conflicts and disputes in the process that wastes resources of time and money and jeopardizes cooperativeness and good working relationships among parties, participants, and stakeholders on projects. The parties, participants, and stakeholders all benefit from the potential DRR holds for construction and related industries, and the practice should be embraced by all in a continuous process improvement of project delivery.

Chapter Review Questions

1) Dispute reputation rating (DRR) is designed to replace the experience modification rating (EMR) in construction. True/False?
2) DRR helps organizations track and improve their dispute-handling reputation. True/False?
3) DRR encourages businesses to rely more on litigation and arbitration. True/False?
4) Dispute recordkeeping is essential for calculating an accurate DRR. True/False?
5) DRR is only applicable to contractors and not to owners or consultants. True/False?
6) DRR can help wean businesses off resource-wasting dispute resolution methods. True/False?
7) DRR is a reactive tool used only after disputes have escalated. True/False?
8) Which of the following is a primary purpose of DRR?
   a) To increase litigation rates
   b) To monitor and rate dispute involvement
   c) To eliminate contracts
   d) To replace insurance policies

9) What type of disputes does DRR aim to track and reduce?
   a) Informal disagreements
   b) Design errors
   c) Negotiated, mediated, arbitrated, and litigated disputes
   d) Budget overruns
10) Which of the following is NOT a benefit of implementing DRR?
   a) Increased reliance on third-party resolution
   b) Improved dispute reputation
   c) Reduced exposure to litigation
   d) Enhanced project competitiveness
11) What is a key component of DRR implementation?
   a) Avoiding documentation
   b) Hiring external arbitrators
   c) Ignoring past disputes
   d) Dispute recordkeeping
12) Which of the following questions is commonly asked during bid evaluations to assess dispute history?
   a) "What is your insurance premium?"
   b) "Have you ever missed a payment?"
   c) "Are there any judgments or claims pending against your organization?"
   d) "Do you use union labor?"
13) DRR is most similar in concept to which of the following industry metrics?
   a) OSHA safety rating
   b) Experience modification rating (EMR)
   c) LEED certification
   d) Project cost index
14) Which of the following best describes the goal of DRR?
   a) To promote litigation
   b) To track dispute frequency and severity
   c) To eliminate contracts
   d) To increase project costs
15) DRR stands for ______ reputation rating.
16) DRR is designed to monitor and rate a business's ______ involvement.
17) A key tool for calculating DRR is accurate and consistent ______.
18) DRR aims to reduce reliance on ______ dispute resolution methods.
19) The concept of DRR is modeled after the ______ used in safety and insurance.
20) DRR helps businesses improve their ______ and competitiveness in the industry.

Exercises

Exercise 15.1 DRR Scorecard Development

Objective:
Create a customized dispute reputation rating (DRR) scorecard for a construction firm.

Instructions:

- Choose a real or hypothetical construction company.
- Review its dispute history over the past five years (real or fictional data).
- Develop a DRR scorecard that includes:
  - Number of disputes (negotiated, mediated, arbitrated, litigated)
  - Severity and cost of each dispute
  - Resolution time
  - Impact on client relationships and repeat business
- Assign weighted scores to each category and calculate an overall DRR.
- Present the scorecard and suggest strategies to improve the firm's DRR over time.

Outcome:
Understand how dispute history affects business reputation and how DRR can be used as a proactive management tool.

Exercise 15.2 Dispute Recordkeeping Audit

Objective:
Evaluate and improve a company's dispute recordkeeping system.

Instructions:

- Assume the role of a dispute analyst for a mid-sized construction firm.
- Conduct a mock audit of the firm's dispute records (real or fictional).
- Assess:
  - Completeness and accuracy of documentation
  - Consistency in tracking dispute types and outcomes
  - Integration with project management systems
  - Accessibility for internal review and external reporting
- Identify gaps and propose a revised recordkeeping protocol aligned with DRR goals.
- Create a checklist or template for future dispute documentation.

Outcome:
Learn how systematic recordkeeping supports dispute prevention and enhances DRR credibility.

Discussion

1) How can a strong dispute reputation rating (DRR) become a competitive advantage in the construction industry? Discuss how owners and clients might use DRR when selecting contractors and how firms can leverage it to win more work.
2) Should dispute records be made publicly available as part of a firm's reputation profile or kept confidential to protect business interests? Debate the balance between transparency, accountability, and competitive sensitivity in dispute recordkeeping.

Index

e